Simon Bühlmann

Accessing the Nanoscale in Ferroelectrics

Simon Bühlmann

Accessing the Nanoscale in Ferroelectrics

Fabrication, Analysis and Modelling of Patterned and Self- Assembled Ferroelectric Nanostructures

Südwestdeutscher Verlag für Hochschulschriften

Impressum/Imprint (nur für Deutschland/ only for Germany)

Bibliografische Information der Deutschen Nationalbibliothek: Die Deutsche Nationalbibliothek verzeichnet diese Publikation in der Deutschen Nationalbibliografie; detaillierte bibliografische Daten sind im Internet über http://dnb.d-nb.de abrufbar.

Alle in diesem Buch genannten Marken und Produktnamen unterliegen warenzeichen-, marken- oder patentrechtlichem Schutz bzw. sind Warenzeichen oder eingetragene Warenzeichen der jeweiligen Inhaber. Die Wiedergabe von Marken, Produktnamen, Gebrauchsnamen, Handelsnamen, Warenbezeichnungen u.s.w. in diesem Werk berechtigt auch ohne besondere Kennzeichnung nicht zu der Annahme, dass solche Namen im Sinne der Warenzeichen- und Markenschutzgesetzgebung als frei zu betrachten wären und daher von jedermann benutzt werden dürften.

Verlag: Südwestdeutscher Verlag für Hochschulschriften Aktiengesellschaft & Co. KG
Dudweiler Landstr. 99, 66123 Saarbrücken, Deutschland
Telefon +49 681 37 20 271-1, Telefax +49 681 37 20 271-0, Email: info@svh-verlag.de
Zugl.: Lausanne, Swiss Federal Institute of Technology, Dissertation, 2004

Herstellung in Deutschland:
Schaltungsdienst Lange o.H.G., Zehrensdorfer Str. 11, D-12277 Berlin
Books on Demand GmbH, Gutenbergring 53, D-22848 Norderstedt
Reha GmbH, Dudweiler Landstr. 99, D- 66123 Saarbrücken
ISBN: 978-3-8381-0564-2

Imprint (only for USA, GB)

Bibliographic information published by the Deutsche Nationalbibliothek: The Deutsche Nationalbibliothek lists this publication in the Deutsche Nationalbibliografie; detailed bibliographic data are available in the Internet at http://dnb.d-nb.de.

Any brand names and product names mentioned in this book are subject to trademark, brand or patent protection and are trademarks or registered trademarks of their respective holders. The use of brand names, product names, common names, trade names, product descriptions etc. even without
a particular marking in this works is in no way to be construed to mean that such names may be regarded as unrestricted in respect of trademark and brand protection legislation and could thus be used by anyone.

Publisher:
Südwestdeutscher Verlag für Hochschulschriften Aktiengesellschaft & Co. KG
Dudweiler Landstr. 99, 66123 Saarbrücken, Germany
Phone +49 681 37 20 271-1, Fax +49 681 37 20 271-0, Email: info@svh-verlag.de

Copyright © 2008 Südwestdeutscher Verlag für Hochschulschriften Aktiengesellschaft & Co. KG and licensors
All rights reserved. Saarbrücken 2008

Produced in USA and UK by:
Lightning Source Inc., 1246 Heil Quaker Blvd., La Vergne, TN 37086, USA
Lightning Source UK Ltd., Chapter House, Pitfield, Kiln Farm, Milton Keynes, MK11 3LW, GB
BookSurge, 7290 B. Investment Drive, North Charleston, SC 29418, USA
ISBN: 978-3-8381-0564-2

*To my parents
and my lovely wife*

Abstract

The continuous downscaling of microelectronic circuits combined with increasing interest in ferroelectric thin films for non-volatile random access memories (FeRAM) is drawing great attention to small ferroelectric thin film structures. There are various challenges and open questions related to processing and theoretical understanding. The processing must assure damage-free ferroelectric capacitors of reproducible properties. More theoretical understanding is necessary to estimate the impact of size effects on the stability of ferroelectric polarization, domain configurations, switching properties, and other properties such as the piezoelectric response.

Pb(Zr,Ti)O$_3$ (PZT) is one of the favorite materials for FeRAMs. Today's capacitors are polycrystalline with sizes in the micron range. Each capacitor is composed of a large number of grains, guaranteeing the averaging of electrical properties. If only a few grains were present, the scattering of properties from one capacitor to another would take place. This can be avoided using oriented single-crystalline capacitors.

An important processing issue is patterning. The switchable polarization might be reduced due to etching damage, a frequently reported current fabrication problem. There is also the question of domain configuration - processing relations. The optimal solution of (001) oriented tetragonal PZT cannot be realized in large capacitors because ferroelastic 90° domains form spontaneously upon cooling through the transition temperature and are stabilized for reasons of stress compensation.

In this work, we investigated two routes to fabricate small ferroelectric structures: A subtractive route and an additive route. Patterning was done by electron beam lithography (EBL) in order to access the sub 100nm range. Dry etching was done in an electron cyclotron resonance (ECR/RF) reactor working at very low pressures and ion energy.

In the **subtractive** route, the starting point were continuous 50 to 200nm thick Pb(Zr$_{0.4}$,Ti$_{0.6}$)O$_3$ films, which were deposited on conductive, Nb-doped SrTiO$_3$ (100) substrates. The films were c-axis oriented with an a-axis fraction of about 20%. The films were patterned by means of EBL. The positive poly-methyl-methacrylate (PMMA) resist was spun on the PZT film, a dot pattern was directly written by the e-beam, and after development a Cr layer was evaporated. The obtained Cr patterns served as hard mask for PZT dry etching. Features with an aspect ratio of up to 2 could be cut out of a 200nm thick PZT film. We have shown that dry etching in an ECR/RF reactor provides damage-free single crystalline sub-200nm features which were still ferroelectric. The smallest features obtained had a lateral size of 80nm. We found that the resolution of the EBL was limited by the backscattering of electrons from the high density PZT layer. Piezoelectric sensitive atomic force microscopy (PAFM) revealed an increase in the piezoelectric response when the feature's aspect ratio was increased. No poling was applied prior to the measurements. The measured piezoelectric coefficient increased by a factor of more than 3 compared to the continuous film. Un-clamping was found to account only for a portion of the increase. We suppose that patterning leads to a domain reconfiguration and unpinning of domains such that a-domains become mobile and contribute to the piezoresponse. This hypothesis is supported by local PAFM measurements on the continuous film. Here, the same behavior was observed when the film underwent local training cycles just before the PAFM measurement. For training, the conductive tip was used to locally apply bipolar voltage pulses well above the coercive field. The piezoresponse at an untrained position was $25 pm/V$ and increased to $160 pm/V$ after training. Theory shows that this corresponds well with the sum of contributions from the clamped film ($105 pm/V$) plus the contribution of mobile a-domains ($60 pm/V$).

It is also important to investigate non-destructive fabrication processes to reduce the negative impact of interface defects and grain boundaries. This was achieved using an **additive** route. The fact, that PZT nucleation on Pt surfaces is difficult but easy on TiO_2 was used for the site controlled growth of PZT single crystals on patterned $2nm$ thick TiO_2 layers serving as affinity spots. This route was carried out on Pt (111) and (100) surfaces. A $50nm$ thick Pt layer was epitaxially grown on single crystalline $SrTiO_3$ (111) and MgO (100) substrates, and then covered by a $2nm$ thick TiO_2 layer. On both substrates, the TiO_2 was found to be (100) oriented. In this route, EBL was used to pattern the TiO_2 layer into seed islands. As only $2nm$ of TiO_2 layer have to be etched, a new type of negative organic mask could be used, which circumvented the lift off process. Direct deposition of $Pb(Zr_{0.4},Ti_{0.6})O_3$ on the seed island covered Pt (111) surfaces led to the nucleation of triangular shaped crystals only on the TiO_2 seeds and not elsewhere. The lateral size of the triangles was between 30 and $150nm$. The uniform orientation of the triangle's sidelines implies epitaxial growth of the PZT. Applying a $1nm$ thick $PbTiO_3$ starting layer prior to PZT deposition increased the nucleation density. A $2nm$ thick $PbTiO_3$ starting layer prior to PZT deposition led to square shaped crystals on small TiO_2 seeds implying the growth of (001) PZT. The squares showed a uniform size distribution, but their in-plane orientation was random. In all cases, the nucleus density was found to be 60 times less on Pt than on the affinity spots. On Pt, triangular crystals have been obtained. The nucleation controlled character was expressed in an observable depletion layer around larger TiO_2 seeds. A theoretical nucleation model was able to well explain this behavior and the relevant parameters such as activation energies of diffusion and desorption could be derived from experiments.

On the Pt (100) surface, direct deposition of PZT led to the nucleation of a non-ferroelectric phase on most parts of the substrate. A depletion around TiO_2 seeds was again observed. PZT crystals were found exclusively on some TiO_2 seeds. PAFM measurements revealed that all PZT crystals obtained by the additive route were ferroelectric. The ferroelectric features have been investigated by means of PAFM. Complex domain structures were identified, the smallest observed domains had diameters of $15nm$. The thinnest investigated feature showing ferroelectricity was $6nm$ thick.

An attempt was made to compare experimental findings with the Landau-Devonshire-Ginzburg phenomenological theory. The $d_{33,f}$ loop together with the CV-loop of a $600nm$ thick, mostly c-oriented film could be fairly well described by the theory. Deviations could be identified as domain wall contributions. The critical thickness for ferroelectricity was estimated using in addition literature data for domain wall energies. The critical thickness was calculated as 1 to $2nm$, which does not contradict experimental findings.

Kurzfassung

Die voranschreitende Verkleinerung mikroelektronischer Bauteile und das zunehmende Interesse an ferroelektrischen Dünnschichten für nichtflüchtige Direktzugriffsspeicher (FeRAM), zieht immer mehr Aufmerksamkeit auf kleine ferroelektrische Dünnschichtstrukturen. Bei sehr kleinen Dimensionen sind Grösseneffekte bezüglich der ferroelektrischen Phasenumwandlung, der Domänenstruktur und der piezoelektrischen Aktivität zu erwarten.

Pb(Zr,Ti)O$_3$ (PZT) ist eines der beliebtesten Materialien für FeRAMs. Die Skalierung zu kleineren Dimensionen ist ein wichtiger Schritt um hochintegrierte Speicher fabrizieren zu können. Heutige ferroelektrische Kondensatoren sind polykristallin und haben eine Grösse im Mikrometerbereich. Jeder einzelne Kondensator wird von einer grossen Anzahl Körner gebildet. Dies garantiert eine einheitliche Mittelwertbildung der elektrischen Eigenschaften. Wenn nur wenige Körner vorhanden sind, führt dies zur unerwünschten Streuung der Eigenschaften. Durch den Gebrauch von einheitlich orientierten Einkristallen kann dies verhindert werden.

Die Skalierung zu kleinsten Dimensionen wirft neue Problemstellungen sowohl bezüglich des Materials und des Herstellungsprozesses, als auch grundlegende theoretische Fragen auf. Die beobachtete Polarisation kann durch den Ätzprozess beeinträchtigt und herabgesetzt werden. Eine wichtige Rolle spielt die Domänenkonfiguration. Das optimale Konzept einer komplett (001) orientierten tetragonalen PZT Schicht kann nicht erreicht werden, da ferroelastische 90° Domänen beim Abkühlen entstehen, welche interne Spannungen abbauen.

In dieser Arbeit verfolgen wir ein subtraktives und ein additives Verfahren, um kleine ferroelektrische Strukturen herzustellen. Um Dimensionen unterhalb von 100nm zu erreichen, wurde die Strukturierung mittels Elektronenstrahllithographie (EBL) durchgeführt. Um Defekte während des Ätzprozesses möglichst klein zu halten, erfolgte er in einem Reaktor, der mittels Elektron-Zyklotron-Resonanz bei sehr tiefem Druck und kleiner Ionenenergie arbeitet.

In der **subtraktiven** Route verwenden wir 50−200nm dicke Pb(Zr$_{0.4}$,Ti$_{0.6}$)O$_3$ Schichten, welche epitaktisch auf elektrisch leitende, Nb dotierte SrTiO$_3$ (100) Einkristallsubstrate abgeschieden wurden. Die Schichten waren bezüglich der c-Achse orientiert. Der Anteil an a-Domänen lag bei etwa 20%. Für die Strukturierung wurde die EBL direkt auf der PZT Schicht durchgeführt. Der verwendete Lack war Poly-Methyl-Methacrylat (PMMA, Plexiglas). Kombiniert mit einem Lift-Off Prozess konnten so Chromstrukturen auf der PZT Schicht fabriziert werden, die während des Ätzprozesses als harte Maske fungierten. Die Auflösung des EBL Prozesses wurde durch die Rückstreuung von Elektronen an der dichten PZT Schicht limitiert. Mit dieser subtraktiven Route konnten aus einer 200nm dicken PZT Schicht Strukturen geätzt werden, die ein Längenverhältnis von 2:1 (Höhe / Breite) aufwiesen. Die kleinste seitliche Ausdehnung betrug 80nm. Es konnte gezeigt werden, dass der Ätzprozess die ferroelektrischen Eigenschaften nicht beeinträchtigt. Piezoelektrische atomkraftmikroskopische (PAFM) Messungen machten deutlich, dass die piezoelektrische Aktivität mit steigendem Längenverhältnis zunahm. Verglichen

IV

mit der piezoelektrischen Aktivität der kontinuierlichen Schicht erhöhte sich der piezoelektrische Koeffizient um mehr als das Dreifache. Entklemmung konnte nur für einen Teil dieser Erhöhung verantwortlich gemacht werden. Als grössten Beitrag zur Erhöhung vermuten wir einerseits eine Änderung der Domänenstruktur von a zu c und andererseits die Befreiung von geklemmten Domänen. Diese Hypothese wird durch PAFM Messungen auf der kontinuierlichen Schicht gestützt, wo dasselbe lokale Verhalten beobachtet wurde, wenn kurz vor der Messung mit der AFM Spitze bipolare Pulse erzeugt wurden, welche das koerzitive Feld überstiegen.

Es ist ebenfalls wichtig, nicht-destruktive Herstellungsprozesse zu erforschen. Dies wurde anhand der **additiven** Route erreicht. Die Tatsache, dass die Nukleation von PZT auf Pt Oberflächen schwierig, auf TiO_2 Oberflächen aber einfach ist, wurde ausgenutzt, um örtlich kontrolliertes Wachstum von PZT Einkristallen auf TiO_2 Keimen zu erzielen. Diese Route wurde auf Pt (111) und (100) Oberflächen durchgeführt. Zu diesem Zweck wurden $50nm$ dicke Pt Schichten epitaktisch auf $SrTiO_3$ (111) oder MgO (100) Einkristallen abgeschieden und mit $2nm$ TiO_2 beschichtet. TiO_2 wuchs sowohl auf Pt (111) als auch auf Pt (100) in der (100) Orientierung auf. In dieser Route wurde EBL verwendet, um die TiO_2 Schicht zu strukturieren. Da nur $2nm$ TiO_2 geätzt werden müssen, konnten wir einen neuen Typus negativen organischen Lacks benutzen, was den Lift Off Prozess überflüssig machte. Auch hier was die Auflösung durch die starke Elektronenrückstreuung der limitierende Faktor. Der Durchmesser der kleinsten TiO_2 Keime betrug $60nm$. Die direkte Beschichtung von $Pb(Zr_{0.4},Ti_{0.6})O_3$ auf Pt (111) Oberflächen führte zur Nukleation von ausschliesslich dreieckigen Kristallen auf den vordefinierten TiO_2 Keimen und nicht anderswo. Die Kantenlänge der Kristalle betrug zwischen 30 und $150nm$. Die einheitliche Orientierung der Kantenlänge bezüglich des Substrates interpretieren wir als epitaktisches Wachstum. Wurde vor der PZT Beschichtung $1nm$ dicke $PbTiO_3$ Schicht aufgetragen, fand die Nukleation auch auf dem Pt statt. Wurden zuvor $2nm$ $PbTiO_3$ aufgetragen, führte dies zu einer Umorientierung der auf den TiO_2 Keimen nukleierten PZT Kristallen: Deren Form war nun quadratisch, was (001) Wachstum impliziert. Auf der Pt Oberfläche konnte das Wachstum von dreieckigen Kristallen beobachtet werden. Die Nukleationsdichte auf Pt war dabei 60 mal kleiner als auf TiO_2. Der nukleationskontrollierte Charakter der PZT Abscheidung äusserte sich einer Verarmung der Keimdichte nahe grossflächigen TiO_2 Keimen. Dieses Verhalten konnte anhand eines theoretischen Nukleationsmodell erklärt, und Parameter wie die Aktivierungsenergie für Diffusion und Desorption bestimmt werden.

Auf der Pt (100) Oberfläche konnte überall die Nukleation einer unbekannten Phase beobachtet werden, und eine Verarmungsregion nahe grosser TiO_2 Keime war ebenfalls sichtbar. PZT Kristalle nukleierten nicht überall, und wenn, dann nur auf TiO_2 Keimen. PAFM Messungen auf PZT Kristallen, die mit der additiven Methode hergestellt wurden zeigten, dass diese ferroelektrisch sind. Vielschichtige Domänenstrukuren wurden ausgemacht, wobei die kleinsten Domänen eine Ausdehnung von $15nm$ aufwiesen. Der dünnste gemessene ferroelektrische Kristall war $6nm$ hoch.

Es wurde der Versuch unternommen, experimentelle Ergebnisse mit der Landau-Devonshire-Ginzburg (LDG) Theroie zu vergleichen. Gemessene $d_{33,f}$ und CV-Hysteresen einer $600nm$ dicken, vorwiegend c-Achse orientierten Schicht, konnten mit guter Übereinstimmung anhand der LDG Theorie beschrieben werden. Abweichungen wurden als Beiträge von Domänenwänden identifiziert. Um die kritische Schichtdicke für Ferroelectrizität abzuschätzen, wurden Daten für Domänenwandenergien aus der Literatur herbeigezogen. Den Berechnungen zufolge liegt diese kritische Dicke zwischen 1 und $2nm$, was mit unseren Messungen nicht im Widerspruch steht.

Contents

1 Introduction   **1**
    1.1 Ferroelectric Materials 1
    1.2 Memory Applications of Ferroelectric Materials 2
         1.2.1 Two Types of Ferroelectric Random Access Memories (FeRAM) 2
         1.2.2 FeRAM Produced to Date 3
         1.2.3 FeRAM vs. Others 4
    1.3 Ferroelectric Thin Films 4
    1.4 Downscaling 5
    1.5 Goal of the Thesis and Approach 5
    1.6 Outline 7

2 Theoretical Background and State of the Art   **9**
    2.1 Fabrication Methods for Small Ferroelectric Cells 9
         2.1.1 Electron Beam Lithography Methods 9
         2.1.2 Focused Ion Beam Etching 10
         2.1.3 Imprint Technique 10
         2.1.4 Selective (Re-)Growth Methods 11
         2.1.5 Random Nucleation 11
    2.2 Size Effects in Ferroelectric Thin Films and Structures 13
         2.2.1 What is Ferroelectricity ? 14
         2.2.2 The Landau Phenomenological Theory 14
                2.2.2.1 Size Effects Due to Depolarizing Fields 15
                2.2.2.2 Surface Effects 17
         2.2.3 Experimental Observations of Size Effects 20
    2.3 Nucleation and Growth of Sputtered Epitaxial Thin Films 23
         2.3.1 General Model 23
         2.3.2 Epitaxy on Oxyde Ceramics: A Phenomenological Approach 25
                2.3.2.1 Coherent, Semicoherent and Incoherent Interfaces 26
                2.3.2.2 Dislocation Formation 27
                2.3.2.3 Metal/Ceramic Interfaces 27
                2.3.2.4 Nb on Al_2O_3 and Nb, Cu on TiO_2 28
                2.3.2.5 Epitaxial Pt on Single Crystalline Substrates 28
    2.4 $Pb(Zr,Ti)O_3$ on Single Crystalline Substrates 30
         2.4.1 Domain Formation 30
         2.4.2 Influence of Film Thickness on Domains 33

		2.4.3	Domain Spacing	34
		2.4.4	The Effect of Seed Layers	35
		2.4.5	Seed Layers for Pb(Zr,Ti)O$_3$	37

3 Growth of Epitaxial Thin Films — 41

- 3.1 Deposition of Thin Films 42
    - 3.1.1 Sputtering 42
- 3.2 X-Ray Diffraction 44
    - 3.2.1 $\theta - 2\theta$ Scans 44
    - 3.2.2 Rocking Curves 44
    - 3.2.3 Pole-Figure Measurement 45
- 3.3 SrTiO$_3$ Substrate Preparation 45
- 3.4 Pb(Zr,Ti)O$_3$ on SrTiO$_3$ (100) 47
    - 3.4.1 X-Ray Characterization 49
    - 3.4.2 d_{33} Interferometer Measurements on the Continuous Film ... 53
    - 3.4.3 Estimations Using the Landau-Ginzburg-Devonshire (LDG) Theory ... 54
        - 3.4.3.1 Estimation of the Coercive Field 55
        - 3.4.3.2 Estimation of the Landau Parameters 56
- 3.5 Pb(Zr,Ti)O$_3$ on SrTiO$_3$ (111) 58
- 3.6 Pt on SrTiO$_3$ (111) 63
- 3.7 Pt on MgO (100) and SrTiO$_3$ (100) with Ir Seeding Layer 63
- 3.8 TiO$_2$ on Pt (100) and Pt (111) 65
- 3.9 Pb(Zr,Ti)O$_3$ Depositions 70
    - 3.9.1 Direct Deposition of Pb(Zr,Ti)O$_3$ on Pt (100) 70
    - 3.9.2 PbTiO$_3$ on Pt (100) 70
    - 3.9.3 Pb(Zr,Ti)O$_3$ on Pt (100) with TiO$_2$ Seed- and PbTiO$_3$ Starting Layer .. 71
    - 3.9.4 Pb(Zr,Ti)O$_3$ on Pt (111) with TiO$_2$ Seed Layer 71
- 3.10 Summary and Conclusion 75
    - 3.10.1 Thin Films for the Subtractive Route 75
    - 3.10.2 Thin Films for the Additive Route 75
        - 3.10.2.1 Deposition of Pt 75
        - 3.10.2.2 Deposition of TiO$_2$ 75
        - 3.10.2.3 Deposition of Pb(Zr,Ti)O$_3$ 76

4 Fabrication of Ferroelectric Nano-Structures — 77

- 4.1 Electron Beam Lithography (EBL) 77
    - 4.1.1 EBL Resists 79
        - 4.1.1.1 Polymethylmethacrylate (PMMA) 79
        - 4.1.1.2 Calixarenes 80
    - 4.1.2 Resolution Limits 81
- 4.2 Experimental Methods for the Lithography Process 84
    - 4.2.1 Preparation of Resist Thin Films for EBL 84
    - 4.2.2 Evaporation 84
    - 4.2.3 Dry Etching in an ECR/RF Reactor 85
- 4.3 The Subtractive Route[1] 88
- 4.4 The Additive Route[2] 93

		4.4.1	Fabrication of TiO$_2$ Dots using PMMA	93

       4.4.1   Fabrication of TiO$_2$ Dots using PMMA 93
       4.4.2   Fabrication of TiO$_2$ Dots using Calixarene[3] 98
       4.4.3   Pb(Zr,Ti)O$_3$ Crystals Obtained on TiO$_2$/Pt/SrTiO$_3$ (111) 101
               4.4.3.1   Pb(Zr,Ti)O$_3$ Deposition without PbTiO$_3$ Starting Layer 101
               4.4.3.2   Pb(Zr,Ti)O$_3$ Deposition with 1nm PbTiO$_3$ Starting Layer . . . 103
               4.4.3.3   Pb(Zr,Ti)O$_3$ Deposition with 2nm PbTiO$_3$ Starting Layer . . . 104
       4.4.4   Pb(Zr,Ti)O$_3$ Crystals Obtained on TiO$_2$/Pt/Ir/MgO (100) 105
       4.4.5   Site Controlled Nucleation and Growth 106
   4.5   Summary and Conclusion 112

5   Piezoelectric Atomic Force Microscopy (PAFM) Measurements   113

   5.1   Introduction to PAFM 114
   5.2   Quantifying the PAFM Piezoelectric Response 115
   5.3   PAFM on Structures Obtained by the Subtractive Route 117
       5.3.1   Measurements on the Continuous Pb(Zr,Ti)O$_3$ Film 117
               5.3.1.1   PAFM Polarization Experiments[4] 117
               5.3.1.2   Nano-Training and Recovery of Ferroelectricity[5] 119
       5.3.2   PAFM Measurements on Patterned Features 122
   5.4   PAFM on Structures Obtained by the Additive Route 130
       5.4.1   PAFM on Features Obtained without PbTiO$_3$ Starting Layer 130
               5.4.1.1   PAFM on Features Obtained on Pt (111)/SrTiO$_3$ (111) 130
               5.4.1.2   PAFM on Features Obtained on Pt/Ir/MgO (100) 135
       5.4.2   PAFM on Features Obtained with PbTiO$_3$ Starting Layers 139
               5.4.2.1   PAFM on Features with 1nm PbTiO$_3$ Starting Layer 139
               5.4.2.2   PAFM on Features Obtained with 2nm PbTiO$_3$ Starting Layer . 141
   5.5   Summary and Conclusion 146

General Conclusions   147

Outlook   149

For comments, suggestions, corrections, write to buehli@buehli.com

Chapter 1

Introduction

1.1 Ferroelectric Materials

Ferroelectricity is a phenomenon of crystalline matter. It may occur in crystals with a point group of one of the ten that allow for a polar axis. Ferroelectricity is thus necessarily accompanied by piezoelectricity and pyroelectricity. In a ferroelectric phase, a spontaneous electrical polarisation develops that is oriented within ferroelectric domains along possible equilibrium directions as defined by the crystal symmetry. The polarization can be reoriented along such directions - and in any case flipped by $180°$ - by applying an electrical field. Ferroelectric materials usually exhibit a para- to ferroelectric phase transition. The spontaneous polarization P is due to a polar distortion of the lattice with respect to the high symmetry paraelectric phase. P is an order parameter of this phase transition and can be quantified because it can be switched and thus measured, and because there is a defined polar distortion of the lattice related to P. Ferroelectrics exhibit many useful properties. The ease to flip polarization leads to outstanding dielectric, piezoelectric and pyroelectric properties. Their piezoelectric constant is up to 1000 times larger than the one of quartz. The large electromechanical coupling gave rise to important applications in ultrasound imaging, acoustic filters, motion and vibration sensors. The large dielectric constants are widely exploited to achieve large capacity densities. The pyroelectric effect is applied in infrared detectors for sensors and imaging. Non-linear behavior in dielectric and electromechanical coupling are exploited to tune capacitors, the refractive index of active optical devices, and the mechanical response of electrostrictive actuators. Semiconducting ferroelectrics exhibit tunable thermistor properties.

The ferroelectric switching as such is exploited only recently in ferroelectric memories. Perovskite type materials are used in this application, such as lead zirconate titanate (PZT), bismuth titanate (BT) and strontium bismuth tantalate (SBT). The bistable nature of ferroelectric switching is used to define the two states of a non-volatile memory. As switching is reversible, ferroelectric materials allow for a new type of non-volatile random accessible memories, so-called FeRAMs. They are essentially a complementary metal oxide semiconductor (CMOS) dynamic RAM (DRAM), the main difference being that they include a ferroelectric capacitor. It took more than 10 years to implement FeRAMs in low-density memories as available on the market at present. Although in theory an ideal memory for mobile phone applications, FeRAMs are lagging behind in integration density to compete with less performing but cheaper memory types of higher density. The delay of high-density FeRAMs will shift their potential introduction to a time when industry will prepare fabrication of chips at the $65 nm$ node. In order to still consider FeRAMs, it must be proven that ferroelectric memories will have outstanding downscaling properties, that they are reliable down to the expected limit of CMOS circuitry at around $30 nm$ design rule, and that they can be reproducibly fabricated at such small dimensions. The present work must be seen in the light of this development. It is tried to identify future fabrication techniques, to explore the limits of the present ones, and to investigate and explain phenomena that appear in ferroelectrics at small scales.

1.2 Memory Applications of Ferroelectric Materials

The ferroelectric bistability can be exploited for building memories, as the electrical polarization vector can be switched by an external field between two distinct stable states up ("1") and down ("0"). In order to switch at low voltages, the ferroelectric material must be necessarily a thin film. A first device employing the memory effect in ferroelectric films was an array for a display with optical read out based on $Bi_4Ti_3O_{12}$ thin films.[6] Switching was slow and the necessary high voltage was unsuitable for applications. Similar as in the case of magnetic hard disk drives, writing and reading of ferroelectric domains on a continuous thin film can be done by means of an appropriate probe head.[7,8] In this case, reading and writing is performed only in a sequential order. The random addressing of individual bits using a word and a bit line is possible in random access memories. Here, the ferroelectric material has necessarily to be patterned into separated capacitors.

1.2.1 Two Types of Ferroelectric Random Access Memories (FeRAM)

A recent type of memory device is the ferroelectric random access memory (FeRAM) incorporating a ferroelectric film as a capacitor to hold data. Most useful for a memory device would be a non volatile memory in CMOS technology exhibiting short read and write times. Such non volatile memories are needed for high-performance digital cameras, digital audio appliances, digital cameras and other multimedia applications. Two versions of non-volatile ferroelectric memories have been studied. Initially, a concept featuring a field effect transistor (FET) with a ferroelectric gate was proposed (FeFET). This device is called a ferroelectric field-effect transistor. It consists of an FET whose gate dielectric is a ferroelectric. This means that the ferroelectric layer is in direct contact with the semiconductor channel. The electric field emanating from the surface of the poled ferroelectric is used to control the conduction properties of the semiconducting channel.[9] Assuming a p-type Si channel and a polarization vector of the ferroelectric, which is directed towards the channel, leads to an accumulation of the channel's electrons near the interface. This reduces the conductance of the p-type channel. If the polarization is upwards, electrons are depleted in the p-type channel near the interface and higher conductance results. Probing the channel's conductivity provides the information about the polarization. Hence, the data readout in an FeFET is non-destructive and needs no reprogramming. Moreover, the FeFET works at a reduced power consumption compared to other devices. However, fatigue and fabrication problems led to a difficult start of this new device. A major problematic issue is the direct deposition of a ferroelectric on Si at high temperatures. This is particularly a problem for lead containing ferroelectric, because lead reacts with Si.

The interest in FeRAM grew again after progress in oxide film deposition in the late 80's and after development of a new concept with destructive read-out. In this case, the ferroelectric memory is similar to a one-transistor DRAM cell where the DRAM dielectric is replaced by

1.2. MEMORY APPLICATIONS OF FERROELECTRIC MATERIALS

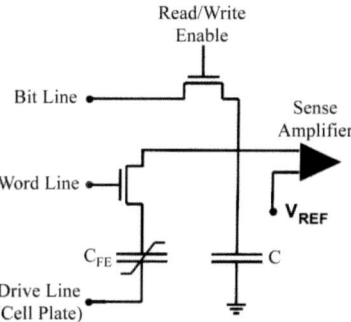

Figure 1.1: Schematic of a 1C1T ferroelectric random access memory cell.[11]

a ferroelectric.[10] A schematic drawing of a 1T1C (1 transistor, 1 capacitor) cell is shown in Figure 1.1, p.3. The ferroelectric material is not situated directly on Si but onto a metallic bottom electrode which serves also as barrier layer to lead diffusion. Like the DRAM, the cell of a FeRAM features a word line and a bit line. The only difference is the drive line present only in the FeRAM cell. To write a positive polarization state, a positive voltage (above the coercive field of the ferroelectric) is applied to the drive line while the bit line is grounded and the word line addressed. To write the inverse polarization, this positive voltage is applied to the bit line and the word line is grounded. To read the information, the bit line is floating (write disabled), and a positive voltage above the coercive field is applied to the drive line while the word line is addressed. This establishes a capacitor divider consisting of C_{FE} and C_{BL} (see Figure 1.1, p.3). Depending on the polarization direction of the ferroelectric, the voltage developed in the bit line can have two different values. Activating the amplifier drives the bit line to the full voltage above the coercive field, and the charge measured provides the information about the polarization state. This reading scheme is destructive, and the initial polarization state has to be written back into the ferroelectric after the read-out.

1.2.2 FeRAM Produced to Date

Nowadays, $256 - Kbit$ chips are mass produced and the fabrication of $128 - Mbit$ FeRAMs are currently in development. The capacitor size is in the μm^2 range, the memory cell size in the range of $10 \mu m^2$. A $32 - Mbit$ chip produced by this technology has a lateral dimension of about $2cm$. Higher integration into $Gbit$ chips for RAM applications requires the downscaling of the memory cell. FeRAMs with lateral capacitor's dimensions of $250nm$ were fabricated by Amanuma et al. (NEC corporation)[12] using Pb(Zr,Ti)O$_3$ on a metal/via stacked plug. Integrated $350 \times 350 nm$ capacitors on W plugs were fabricated by Texas Instruments.[13] The $90nm$ thick Pb(Zr,Ti)O$_3$ capacitors have low voltage switching properties with polarization saturation smaller than $1.8V$. The remanent polarization of the smallest capacitor was $25 \mu C/cm^2$.

Property	SRAM	FLASH	DRAM	FeRAM (projected)
Min. Voltage	$> 0.5V$	$> 12V (\pm 6V)$	$> 1V$	$> 1V$
Write Time	$< 10ns$	$100\mu s/1s$	$< 20nm$	$< 20ns$
Write Endurance	$> 15^{15}$	$< 10^5$	$> 10^{15}$	$> 15^{15}$
Read Time	$< 10ns$	$20ns$	$< 20ns$	$< 20ns$
Read Endurance	$> 10^{15}$	$> 10^{15}$	$> 10^{15}$	$> 10^{15}$
Nonvolatile ?	no	yes	no	yes
Cell Size	$80F^2$	$8F^2$	$8F^2$	$15F^2$

Table 1.1: Comparison of different memory technologies.[11] F designates the half metal pitch.

1.2.3 FeRAM vs. Others

Fast write, low voltage write, low power operation, and high write endurance are the major advantages of ferroelectric memories. FeRAMs offer many benefits over classic non-volatile memories like FLASH or EEPROM.[14] Compared to DRAMs, the FeRAM has about the same properties, but it is non-volatile. Nevertheless, today's FeRAM are still slower than SRAM and their density is smaller than that of available FLASH or DRAMs.[15] Low density FeRAMs are used in FLASH memories which are produced in credit card format. Here, the full advantages of the non-volatile charachter of FeRAM are applied: The electrical power is provided by an *RF* antenna only when needed. The memory is maintained by the FeRAM in the off mode. Table 1.1, p.4 compares the different types of RAMs.

1.3 Ferroelectric Thin Films

Probably the most difficult demands on ferroelectric films are placed on the ferroelectric non-volatile memory structures. The low drive power requirements impose a low film thickness but the films should also hand exhibit near-bulk properties.[16] The temperature at which the ferroelectric material is deposited is usually above the stability of the ferroelectric phase. The ferroelectric crystal structure is created when cooling below the transition temperature. Stress is generated due to lattice mismatch and different thermal expansion coefficients between the substrate and the film. In pseudo-cubic crystals, such as $Pb(Zr,Ti)O_3$, stress is partially released by the formation of $90°$ ferroelectric domains which have the polarization vector in the plane. The presence of a-domains and the clamping of the substrate reduce the piezoelectric response of a thin film.

Initially, ferroelectric KNO_3 was supposed to be a candidate for memory applications,[17,18] but soon $Pb(Zr,Ti)O_3$ became the most prominent material, not only due to its high remanent polarization (about $100\mu C/cm^2$), but also due to its relatively simple perovskite structure and phase diagram, and the low processing temperature of $550 - 650°C$. The main drawback of $Pb(Zr,Ti)O_3$ is its early fatigue on Pt electrodes. Many research efforts were triggered after the finding, that bismuth compounds like bismuth titanate (BT) or strontium bismuth tantalate (SBT, $SrBi_2Ta_2O_9$) show excellent fatigue behavior on Pt.[19–22] Moreover, bismuth compounds switch at low coercive field. Promising results were obtained on structured[21,22] and unstructured SBT,[20] and no size effects were observed down to $1\mu m$.[20,23] However, the main drawback of

1.4. DOWNSCALING

bismuth compounds are the low switching charge ($10\mu C/cm^2$),[24] high processing temperatures of up to $800°C$, the segregation of Bi to the surface, and the difficulty to orient the polar axis. Moreover, in nowadays FeRAMs, Pt is replaced by oxide electrodes (above all IrO_2, but also RuO_2, $SrRuO_2$ and $YBa_2Cu_3O_7$ have been explored) on which $Pb(Zr,Ti)O_3$ shows excellent fatigue behavior.

1.4 Downscaling

High density integration of FeRAMs can only be achieved if the ferroelectric capacitor's size is downsized to some $100nm$. In this range, it is theoretically predicted that the ferroelectric phase becomes gradually unstable.[25] Experimentally, a change in the domain configuration was observed in fine grained ceramics[26] and thin films.[27] This leads to size dependent physical properties, or so called size effects. For the FeRAM, a capacitor size over $700nm$ is too large for general cost-effective applications.[13] Even for the densest design[11,15,28], the lateral size of a $1GBit$ memory cell should not exceed $150 \times 150nm^2$. This implies ferroelectric capacitors having lateral dimensions below $100nm$.[29,30] Compared to the size of a DRAM cell ($130nm$), the smallest FeRAM cell is still $350nm$ large. FeRAM technology is 4 to 8 years behind DRAM technology!
A big advantage of ferroelectrics is the flat domain wall (few lattices in $Pb(Zr,Ti)O_3$ 90° domains, and < 1 lattice in 180° domains) which assures low crosstalk. Theoretically, downsizing to less than $10nm$ is possible. A density of $1.5 Tbit/inch^2$ has recently been achieved on a continuous film with narrowly spaced $12nm$ dots by scanning nonlinear dielectric microscopy (SNDM).[8] The integration in CMOS technology[10,28,31] needs structuring of the ferroelectric layer. At small dimensions, surfaces and interfaces become more and more important. Intermediate small ($1\mu m$) ferroelectric structures have already been studied[19] and important issues such as fatigue,[17,32,33] imprint,[34] optimum film thickness,[30] switching kinetics,[18,35,36] retention of the stored information[37] and size effects[38] have been addressed. In $Pb(Zr,Ti)O_3$ capacitors, an unexplained increase in the maximum and the remanent polarization was observed when the lateral size of the capacitor size decreased to $1\mu m$.[12] An asymmetry in swtiching and size dependent switching patterns were observed on micrometer sized (111) oriented Ca-, Sr-, and La-doped $Pb(Zr_{0.4},Ti_{0.6})O_3$ on Pt (111).[39]
In micrometer-sized polycrystalline capacitors, averaging over a large number of grains guarantees uniform properties. Downscaling to small dimensions leads to a smaller number of grains per capacitor and this in turn leads to scattering of properties. This can be avoided using oriented or single crystalline materials. In single crystalline materials, grain boundaries are avoided. It is expected, that such materials show better properties.[40,41] In order to fabricate FETs, many attempts have been undertaken to deposit epitaxial ferroelectric layers directly on Si to use it as a gate dielectric. Unfortunately, Si easily forms unwanted compounds with the ferroelectric material at the interface. Recently, the epitaxial growth of Pt was achieved on Si (100).[42] Even thought the semiconducting properties of the Si became insufficient in this case, it is nevertheless an important step towards integrating an FET directly on Si.

1.5 Goal of the Thesis and Approach

The goal of this thesis is the fabrication of small ferroelectric $Pb(Zr,Ti)O_3$ structures by subtractive and additive methods, and to study their ferroelectric properties. Size dependent properties

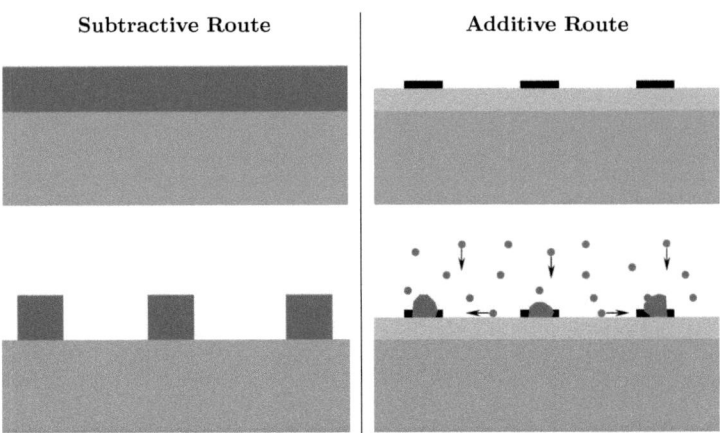

Table 1.2: Schematic of the two fabrication routes.

have already been observed in fine grained ceramics and thin films. This work proposes to study individual small single crystal ferroelectric grains. To simplify the interpretation of results, integration on silicon is avoided, and instead, epitaxial thin films on single crystalline substrates are the basis of this work. A subtractive and an additive route will be studied (see Figure 1.2, p.6). In the subtractive route, Pb(Zr,Ti)O_3 thin films are epitaxially deposited on conductive Nb-doped SrTiO_3 (100) single crystals, and patterned into small features. This involves the (destructive) etching of the ferroelectric layer. In the additive route small Pb(Zr,Ti)O_3 crystals are "naturally" grown on predefined TiO_2 seed sites. Here, an thin TiO_2 seed layer on Pt is patterned instead of the ferroelectric layer. Subsequent Pb(Zr,Ti)O_3 sputter deposition leads to the growth of small Pb(Zr,Ti)O_3 crystals on the seed sites and not elsewhere. The governing mechanism is the very different nature of Pb(Zr,Ti)O_3 nucleation on platinum and on TiO_2. The low chemical affinity of platinum to oxygen leads to long free mean paths before adsorption and together with the high volatility of PbO to a high desorption rate. TiO_2 plays the role of nucleation islands where every incoming species is absorbed immediately. Depositions on two surfaces will be investigated: on Pt (111)/SrTiO_3 (111) and Pt (100)/MgO (100).

To access small dimensions ($< 100nm$), structuring is done by means of electron beam lithography (EBL). In the case of the subtractive route, Cr hard masking with the positive PMMA resist was used to etch the Pb(Zr,Ti)O_3. For the patterning of the $2nm$ thin TiO_2 layer in the additive route, PMMA was also used, but considering the small thickness ($2nm$) of TiO_2 a negative organic Calixarene resist could be used as well. For the characterization of the small ferroelectric features, the local piezoelectric vibration is detected locally with piezoelectric sensitive atomic force microscope (PAFM) using a conductive tip as a mobile top electrode in contact with the Pb(Zr,Ti)O_3 crystals. An *AC* signal applied between the tip and the bottom electrode induces local piezoelectric vibration which is sensed by the tip. This method measures the local amplitude of piezoelectric vibration of the ferroelectric layer and the phase (up or down).

1.6 Outline

Chapter 2 covers the theoretical background and the state of the art. The different fabrication methods for small structures and theoretical considerations concerning size effects in ferroelectrics and their observation are discussed. The nucleation and growth of thin films is presented with emphasis on epitaxial depositions and surface diffusion. Regarding the growth of ferroelectric films, and especially $Pb(Zr,Ti)O_3$, domain formation and the influence of seed layers are discussed.

Chapter 3 presents the characterization of the obtained epitaxial thin films used in this work. The crystallographic orientation of the films is shown, and for $Pb(Zr,Ti)O_3$ ferroelectric measurements are shown.

After an introduction to electron beam lithography (EBL) in Chapter 4, the fabrication of small ferroelectric features obtained by additive and subtractive routes is shown and discussed related to the e-beam process.

Finally, Chapter 5 shows the ferroelectric characterization of the of obtained features. For characterization, piezoelectric atomic force microscopy (PAFM) was used.

General conclusions and the outlook can be found at the end.

Chapter 2

Theoretical Background and State of the Art

2.1 Fabrication Methods for Small Ferroelectric Cells

High density memories require ferroelectric capacitors with lateral dimensions of less than $100nm$.[43] At such small dimension, size effects are supposed to play a significant role: Possible examples are monodomain configurations, which are energetically favourable for very small structures, and depolarization fields causing the instability of the polar phase.[29] There is a strong interest in investigating size effects in ferroelectric cells. Surface effects play a crucial role due to the increasing surface to volume ratio when decreasing the feature size. The required small dimensions for the observation of size effects are difficult to reach with conventional lithography methods, and other techniques are required. Many experiments, which are discussed below, were performed on single crystalline substrates to assure uniform orientation of the polarization and to avoid scattering of properties due to a small number of grains per capacitor at very small dimensions. The obtained epitaxial material also shows better fatigue characteristics.[41] Experimentally, several approaches have been tested to reach the small sizes. Generally, one can distinguish between methods which provide the control over the position of the small ferroelectric cell and those that do not provide this kind of control. In the former case there are electron beam lithography methods (this work and and the group of Alexe et al.), self-assembly through nucleation site controlled growth (this work) and focused ion beam etching (the group of Ramesh et al.). The latter case covers all kind of random self-assembly (the group of Waser and Alexe et al.).

2.1.1 Electron Beam Lithography Methods

Metalorganic layers were used as a negative precursor resist in electron beam lithography (EBL) for the fabrication of small cells of $Pb(Zr,Ti)O_3$ and SBT.[43–48] During irradiation, chemical reactions are locally induced in the metalorganic thin film with an electron beam. After development the patterns have to be pyrolized at $350°C$ and fired at $850°C$ to produce the ferroelectric material of unknown orientation. The $PbZr_{0.3}Ti_{0.7}O_3$ patterns had lateral sizes down to $100nm$. The initial polarization was down. The remanent piezoelectric and the coercive field were independent of lateral size. PAFM measurements of the piezoelectric coefficient showed a value of

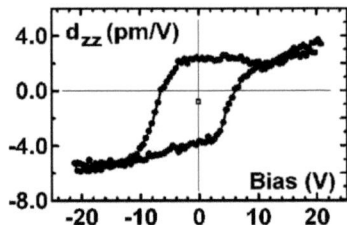

Figure 2.1: Typical loop acquired by PAFM on a $500 \times 500 nm$ cell of Pb(Zr,Ti)O$_3$. The shift is explained by pinning of domains at the free lateral surface.[45]

$d_{33} = 2 - 4 pm/V$ for $100 nm$ cells at maximum applied field. The loops were asymetric (shift downwards) and the coercive field showed a shift to smaller values for features of $1 \mu m$ (see Figure 2.1). The shift increased with decreasing feature size. This was attributed to a polarization imprint due to domain pinning at the free lateral surface and the ferroelectric-bottom electrode interface. Some loops showed a decreased d_{33} after reversing the sign of DC field.

2.1.2 Focused Ion Beam Etching

Single capacitors of Pb(Zr,Ti)O$_3$ with lateral size of $200 nm$ were obtained by focused ion beam milling[49,50] from a $300 nm$ thick Pb(Zr,Ti)O$_3$ film. Pb(Zr$_{0.2}$Ti$_{0.8}$)O$_3$ was epitaxially deposited on $50 nm$ La$_{0.5}$Sr$_{0.5}$CoO (LSCO)/SrTiO$_3$ (100) by PLD at $650°C$ and $500 mT$ oxygen pressure. The top electrode was contacted with a conductive AFM tip featuring a PtIr coating. The parasitic capacitance was in the same order of magnitude for sizes below $1 \mu m$ and had to be subtracted (by an open circuit measurement plus correction due to increased distance to the bottom electrode). The measured polarization charge of a $200 nm$ thick capacitor was $70 \mu C/cm^2$.[49] No size effects could be observed to sizes down to lateral sizes of $400 nm$. Below, effects were related to parasitic capacitances.

The dependence of internal stress on polarization was investigated[51] and an increase in spontaneous polarization was observed when the epitaxial film's thickness of PbZr$_{0.8}$Ti$_{0.2}$O$_3$ on LiAlO$_3$ (001) with LSCO electrodes was deceased. This was attributed to an increase of the observed c-axis parameter due to internal stress arising from lattice mismatch. An "unconstrained" situation was obtained in Pb$_{1.0}$(Nb$_{0.04}$Zr$_{0.28}$Ti$_{0.68}$)O$_3$ when cutting out a laterally $250 nm$ large feature by a FIB.[51] The value of d_{33} doubled.

2.1.3 Imprint Technique

Imprint techniques involving etching of a precursor gel, pyrolyzing and annealing, were used to produce mesoscopic Pb(Zr,Ti)O$_3$ structures.[52] Cells of $300 nm$ lateral size and $50 nm$ height were obtained. The average grain size was $35 nm$. An as fabricated down polarization was observed. The large cells had the same response as the small ones on Si/SiO$_2$/Pt. PAFM loops showed values of $-4 - +6 pm/V$ on Pt and $-20 - +10 pm/V$ on Nb-SrTiO$_3$. The features on SrTiO$_3$ showed a polarization imprint towards the bottom electrode.

2.1. FABRICATION METHODS FOR SMALL FERROELECTRIC CELLS

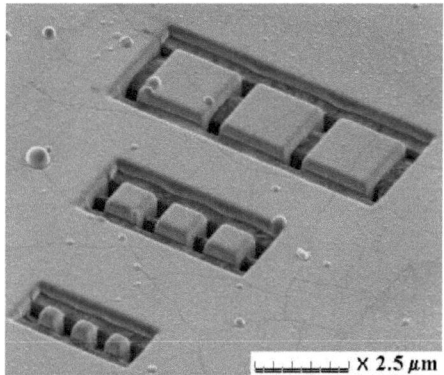

Figure 2.2: Pb(Zr,Ti)O$_3$ features cut by means of a focused ion beam.[50]

2.1.4 Selective (Re-)Growth Methods

Lee et al.[53–55] produced large Pb(Zr,Ti)O$_3$ seed islands for Pb(Zr,Ti)O$_3$ regrowth. A 100nm thick perovskite Pb(Zr,Ti)O$_3$ seed layer was fabricated by sol gel, fired at 700°C on Pt/SiO$_2$/Si and patterned by wet etching into large 13 × 13μm islands figuring as templates.
Then a Pb(Zr$_{0.65}$Ti$_{0.35}$)O$_3$ layer of 200nm was sputter deposited at only 350°C. Pb(Zr,Ti)O$_3$ deposited on bare Pt grew in the pyrochlore phase, and was partially perovskite on the seed islands. To fully transform the Pb(Zr,Ti)O$_3$ into perovskite, an anneal at 700°C was necessary on bare Pt against 540° on the seed layer.[54] The films crystallized in the mixed (101) and (100) orientation. Annealing led to the lateral growth of the perovskite phase circularly around the seed islands into the pyrochlore phase as a function of annealing temperature and time. A saturation in growth distance was found which was found to be dependent on the temperature but not on the annealing time. The reason for the this self-limiting behavior of lateral growth is supposed to be the interface energy.[55] The self-limiting effect in growth was also found in random nucleation. It is worth noting here that selective growth has been reported on a different combination of materials, namely of GaN on AlN seed layers on SiC substrates.[56] GaN nucleated only on AlN seed layer but not elsewhere.

2.1.5 Random Nucleation

Hiboux et al.[57] investigated the early stages of Pb(Zr,Ti)O$_3$ nucleation on (111) oriented Pt using a sputtering process with a Ti and a Pb target. The morphology was investigated as a function of the molecular flux between PbO and TiO$_2$. Triangular crystals down to 15nm were obtained at a PbO flux fraction of 0.55, see (Figure 2.3(a), p.12). TEM analysis showed that the triangular shaped crystals are perovskite. Similar results were reported for the deposition of Pb(Zr,Ti)O$_3$ on Si/SiO$_2$-Pt by MOCVD.[58] A good way to investigate size effects in ferroelectrics is to produce single-crystal, defect-free, monodomain nanostructures with controlled size and orientation. BaTiO$_3$ was deposited by PLD on SrTiO$_3$. Island growth led to 30 − 40nm large and 15nm high crystals randomly distributed on rough steps of SrTiO$_3$.[29] To reach the

CHAPTER 2. THEORETICAL BACKGROUND AND STATE OF THE ART

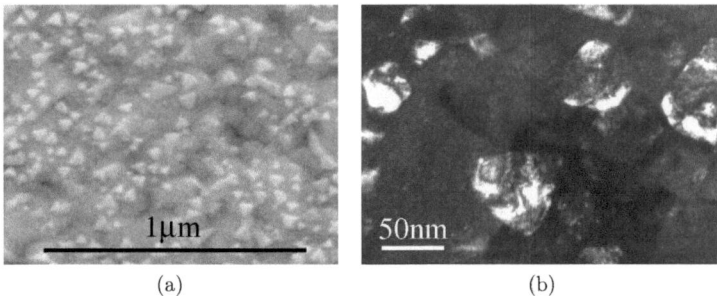

(a)   (b)

Figure 2.3: $PbTiO_3$ seed layer of an equivalent TiO_2 thickness of $5nm$. The lead flux fraction was 0.55. (a) SEM image showing triangular shaped crystals with sizes down to $15nm$. (b) Dark field TEM plane view image using a diffracted beam corresponding to $PbTiO_3$ (111) orientation.[57]

limit of polarization in $PbTiO_3$, Roelofs et al.[59] deposited a very thin film of chemical solution. They achieved randomly distrubuted $PbTiO_3$ grains on Pt in the range of $20-200nm$. PAFM studies on the larger grains showed stripes with widths between 12 and $80nm$ that appeared as a consequence of the appearance of $90°$ domains. A $PbTiO_3$ grain with a diameter of about $18nm$ and a height of $5nm$ didn't show any piezoelectric response, and it was concluded that the size of this grain was below the critical size for ferroelectricity. Unfortunately, no piezoelectric loop on that grain was shown to underline this finding.
Another method to fabricate small particles uses the instability of ultrathin films during high-temperature treatments. After high-temperature annealing, the as-deposited $PbZr_{0.52}Ti_{0.48}O_3$ film breaks into islands with a narrow size distribution.[60] The driving force is the excess of the total free energy of the continuous film compared with a film partially covering the substrate. The free energy is minimized by formation of islands, which lowers the interface area and consequently the interface energy. The obtained island were $20-200nm$ of lateral size. Islands with a height of $9nm$ and a lateral dimension of $50nm$ were also observed. TEM showed epitaxial nanostructures with facets which consist of $\{111\}$ or $\{110\}$ and $\{100\}$ faces. A measured loop proved ferroelectricity and showed a shift upwards.
An interesting mechanism of self-assembly is the segregation of Bi to the surface at elevated temperatures. These randomly distributed metallic features were used as top electrodes on c-axis oriented $Bi_4Ti_3O_{12}$ which was epitaxially grown on $SrTiO_3$ (001) by PLD.[30,61,62] Cooled down in vacuum, the formed segregate shows metallic conduction at room temperature. The segregation forms square shaped depositions on the surface of the ferroelectric layer which were randomly distributed (see Figure 2.4, p.13). The switching of such capacitors was shown by piezoelectric AFM.[62] Top electrodes with lateral sizes of $180nm$ were obtained in that way. Due to slight Bi excess in the film, polarization loops were not saturated. The polarization charge was $10 \mu C/cm^2$. Not all of the cells were found to have the same piezoresponse, most didn't have any piezoresponse at all. The switching of the cells could be shown by piezoelectric AFM.[62]
To address size effects, individual small grains of $SrBi_2Ta_2O_9$ (SBT) and $Bi_4Ti_3O_{12}$ (BT) were investigated by PAFM on polycrystalline films.[63] $150nm$ SBT and BT was grown by PLD on epitaxial $LaNiO_3$ (LNO)/CeO_2/YSZ (yttrium stabilized zirconia)/Si (100). The $150nm$ LNO served as bottom electrode. The grain size of SBT films was $200nm$. So called in-field (DC

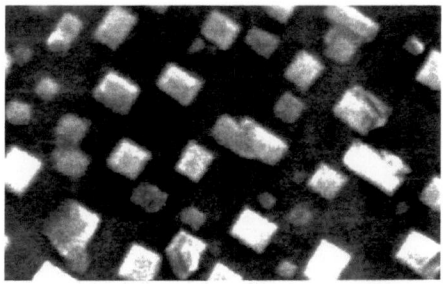

Figure 2.4: Self assembled cubic-shaped electrodes of conductive bismuth compound on top of a $200nm$ thick $Bi_4Ti_3O_{12}$ film. During cooling, Bi is "sweat out" of the film and forms a conductive compound on the surface which can be used as top electrode.[61] The film was deposited by $LSCO/CeO_2/YSZ/Si$.

and simultaneous small AC field, like a small signal response measurement) and remanent loops (DC pulse + waiting time + measurement) were measured by PAFM. At high electric fields, in-field loops showed a piezoresponse which was proportional to the field. This was attributed to electrostriction. The electrostriction coefficient was estimated using equation 3.7, p.56 as $3 \times 10^{-2} m^4 C^{-2}$ in SBT and $7.7 \times 10^{-3} m^4 C^{-2}$ in BT. The saturated d_{33} was measured to be $2pm/V$. On $Pb(Zr,Ti)O_3$ a decrease in the d_{33} was observed for high fields in remanent loop measurements.[63] This was attributed to stress induced depolarization. $200nm$ large SBT grains were still ferroelectric. Finally, Luo et al.[64] fabricated $PbZr_{0.52}Ti_{0.48}O_3$ and $BaTiO_3$ nanoshells exhibiting piezoelectric hysteresis and ferroelectric properties. AFM measurements revealed a d_{33} of $100pm/V$. The loop showed a lower d_{33} value after reversing the DC.

2.2 Size Effects in Ferroelectric Thin Films and Structures

It is well known that the properties of bulk ceramics depend very much on grain size.[26,65-70] The reason is that grain size affects domain configuration. For thin films it is expected that ferroelectricity can't exist any more below a critical film thickness. The effect of depolarizing fields (and strain) can be reduced by domain formation of opposite polarization[71] (or by formation of 90° domains). The gain in energy from the creation of 180° and ferroelastic domains is counterbalanced by the energy of the domain walls. Thus, one can imagine that at small sizes, a domain configuration with few domains, or even a monodomain state[72,73] is favorable to reduce the energy of domain walls.

The existence of size effects has two main causes: Spatial correlation of the ferroelectric polarization and boundary condition of the polarization on the electrodes.[74] With metallic electrodes, the polarization has to drop to zero, with oxide electrodes the polarization does not decrease to zero (free boundary conditions). To understand more about size effect in ferroelectric material, it is useful to address the question about the origin of ferroelectricity.

2.2.1 What is Ferroelectricity ?

Ferroelectricity develops through phase transformation at the Curie temperature T_0 from a high temperature high symmetry phase to a low temperature lower symmetry phase. Ferroelectricity is a collective phenomena involving the cooperation of polar distortions through both, short-range chemical and long range dipolar interactions. The parallel alignment of dipoles in a ferroelectric is due to a long range dipolar interaction in competition with local forces, which prefer a nonpolar state.[75] The correlation length for the polar interaction is estimated to be $10 - 50nm$ along the polar axis and $1 - 5nm$ normal to it.[71,76] If the dimensions of the crystal fall below these values, a change in stability of the ferroelectric phase can be expected.[75] Polarization becomes unstable because the grain size has been reduced below the correlation length necessary for the dipolar interaction to stabilize the ferroelectric phase.

In the example of $PbTiO_3$, the ferroelectric transition is driven by the freezing-in of an unstable phonon of the cubic perovskite structure, where the positively charged metal ions move against the negatively charged oxygens, accompanied by a corresponding change in shape of the unit cell. The oxygen octahedra are almost undistorted, and the relative displacement of the Pb atoms is 0.35Å. The instabilities which lead to ferroelectricity are dominated by large Pb displacements[77] and by the intrinsic instability of the Ti-O chain. The energy of the uniform distortion has a "double well" form. Waghmare et al.[78] calculated the transition temperature by first principle mechanisms for bulk $PbTiO_3$ and found $660K$, which is in acceptable agreement with the measured value $763K$. Ghosez et al.[79] studied theoretically the ground state polarization of stress-free $PbTiO_3$ (001) thin films under short circuited boundary conditions (no depolarization) using quantum mechanical first principles approaches.[80] They predict a ferroelectric ground state down to a thickness of three unit cells with significant enhancement of the polarization at the surface.[79] Tybell et al.[81] investigated experimentally ultrathin sputtered, (001) oriented $PbZr_{0.2}Ti_{0.2}O_3$ films on conductive, Nb doped $SrTiO_3$. A conductive AFM tip was used to inversely pole the layer on a line. The inverse polar state could be measured on films as thin as $4nm$. These tip-poled areas were stable for at least $180h$.

Two main reasons for size effects are addressed in the literature: High depolarizing fields which destabilize the spontaneous polarization and surface effects which become more and more important at small dimensions. Theoretical considerations use the Landau phenomenological theory to explain and predict size effects.

2.2.2 The Landau Phenomenological Theory

The existence of a ferroelectric phase can energetically be explained with a free energy function having two minima (the "double well", see Figure 2.5, p.15). This function can be approximated by a Taylor series in P having only even power numbers (invariance under the symmetry group of the high-symmerty, high-temperature phase, which includes the inversion symmetry).[82] It is a phenomenological description going back to Landau's mean field theory.[83] Depending on the extensions, it is also called the Landau-Devonshire-Ginzburg (LDG) theory. The complete free energy density for ferroelectrics including the polarization P and the strain x was formulated by Devonshire and Ginzburg.[84] The polarization is taken as the order parameter. The necessary constants can be measured experimentally and good agreement was found with the LDG theory. The theory is flexible and allows to include effects thought to influence the stability of the ferroelectric phase at small sizes, such as depolarizing fields, surface layers, misfit strain, domain structure, the anisotropy of the free energy, and Schottky interface layer. Recently, the theory was applied to calculate phase diagrams of single-domain epitaxial $Pb(Zr,Ti)O_3$ films[85] and

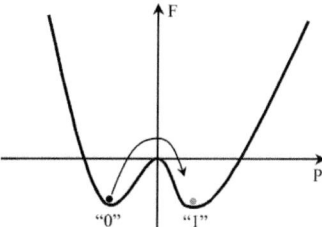

Figure 2.5: Schematic of the free energy as a function of the order parameter (polarization) showing the "0" and the "1" state and the energy barrier to overcome for switching.

switching.[86] Watanabe et al.[38] extended the theory to materials with finite band gap (not perfect insulators) and included defects as stabilizing elements of the ferroelectric phase. The theory can be applied to first and second order transitions for all possible geometries (thin films, spherical particles and edge defined cells). It can predict the critical size at which the polarization vanishes, i.e. at which the Curie temperature drops to $0K$, but also non-linearity of the piezoresponse with field strength.[87] A general form of the Landau free energy term for a free body (or free Gibbs function expansion) is given below. It includes depolarizing fields as well as the influence of surface ordering.

$$F = \int_{(V)} \left[\underbrace{\frac{1}{2}AP^2 + \frac{1}{4}BP^4 + \frac{1}{6}CP^6}_{Landau} + \overbrace{\frac{1}{2}D(\nabla P)^2}^{polarization\ gradient} \underbrace{- E_{dep} P}_{depolarization} \right] dV + \int_{(S)} \underbrace{\left[\frac{1}{2} D\, \delta^{-1} P^2 \right]}_{surface\ term} dS$$

A, B, C are the coefficients in the free Gibbs expansion and can be measured experimentally, T_0 is the Curie temperature of the bulk, and δ is the extrapolation length describing the surface. A is defined as $A = A_0(T - T_0)$. The Landau parameters $A - C$ have been determined for BaTiO$_3$, PbTiO$_3$, and Pb(Zr,Ti)O$_3$ between 1950 and 1970 and are still used in recent literature. For BaTiO$_3$, the values from Mitsui et al.[88] and Merz[89] are used, and the work from Zhirnov[90] is sometimes used to estimate the A parameter. For PbTiO$_3$, the values from Remeika et al.[91] and Haun et al.[92] are used. The work from Haun et al.[92] is also used for Pb(Zr,Ti)O$_3$. A second order transition is obtained with positive B, while a first order transition is obtained when $B < 0$ and $C > 0$ is needed to stabilize the polarization.[93] D are gradient coefficients which measure the delocalized coupling strength of the dipoles. The values of D were determined by Zhirnov[90] and Wemple.[76]

2.2.2.1 Size Effects Due to Depolarizing Fields

As films become thinner, depolarization field effects become important, lowering both, T_0 and the spontaneous polarization P_s. This was predicted by Binder and Tilley et al.[94,95] For sufficiently thin films (or small particles) both reach zero, and ferroelectricity can become impossible.[96] The idea that depolarizing fields can inhibit the spontaneous polarization has been proposed as early as 1950.[73,97] Huge depolarizing fields can theoretically be predicted. Considering an epitaxial

Pb(Zr$_{0.4}$,Ti$_{0.6}$)O$_3$ film on SrTiO$_3$ (100) the depolarizing field $E_{dep} = \frac{P}{\epsilon}$ is $4300 kV/cm$ which is 25 times bigger than the switching field (the values at zero field were taken from Figure 3.12(b), p.54 and Figure 3.13(a), p.55). The presence of compensation charges is required for material stability.[98] Even though the compensation charges can closely cancel the depolarization field, an existing residual depolarization could still be sufficiently large to lead to observable effects, above all as the film thickness shrinks. For small enough dimensions, the electrostatic energy of the depolarizing field will be bigger than the polarization energy and spontaneous polarization would not develop. Jaccard et al.[73] compared the polarization energy of ferroelectric particles having short circuited surfaces with the energy of isolated non conductive particles where strong depolarization fields are expected. Large crystals can reduce the depolarizing field by domain formation. However, the creation of a domain wall is energy consuming and one can expect that domain formation is unfavorable below a critical size. Below that size, the inhibition of the polarization can be attributed only to depolarizing fields. This was found in electrically isolated KH$_2$PO$_4$ particles which did not show the splitting of the tetragonal a and c axis (the polar distortion) in X-ray measurements when the size was smaller then $150nm$. In the electrically conductive medium, the particles deformed into the tetragonal state even for the smallest reached sizes of $50nm$ and showed normal spontaneous polarization.[73]

The effect of incomplete compensation at semiconducting interfaces was theoretically and experimentally investigated by Batra et al.[93,98–100] using an asymmetric arrangement of ferroelectric triglycine sulphate (TGS) sandwiched between a metallic and a semiconducting electrode under short circuited conditions. The compensation charge distribution is asymmetric and consequently depolarization fields cannot be sufficiently reduced by domain formation. The influence of depolarizing fields was evidenced comparing the situation in a dark and in illuminated environment (where charge compensation takes place even on the semiconductor side, see Figure 2.6(a), p.17). In the dark, existing band bending between the ferroelectric and the semiconducting electrode gives rise to a depletion layer in the semiconductor (the depletion layer in the ferroelectric is not considered). This leads to non complete compensation of polarization charges at the interface and hence to a depolarizing field which destabilizes the polar state at a critical thickness. For a TGS film between semiconducting electrodes and heated to $45°C$, the spontaneous polarization was calculated as a function of the film thickness (see Figure 2.6(a), p.17). At high carrier concentration (lower band gap) the critical thickness was lower ($700nm$) than for higher band gap semiconductor electrodes ($1200nm$). Between metallic electrodes, the critical thickness of TGE was calculated to $400nm$ (see Figure 2.6(b), p.17). At any temperature the corresponding value of the polarization is lower than the bulk value. It is shown that the transition temperature drop causes the polarization to drop to zero below a critical film thickness and the depolarizing field increases with decreasing film thickness, passes through a maximum and drops also to zero due to inhibition of the ferroelectric phase.

The effect of the depolarization in small BaTiO$_3$ particles was studied by Shih et al.[101] Considering the crystal to break up into 180° domains, the depolarization energy and the domain wall energy was considered in the Landau model. A Schottky space charge layer beneath the surface accounted for the non-perfect insulating properties. Such space charge layers have been observed experimentally in single crystal BaTiO$_3$ by Triebwasser.[102] These $10 - 1000nm$ thick layers may result from high oxygen vacancy concentration at the surface.[103] The calculations show that the ferroelectric transition temperature of small particles can be substantially lower than that of the bulk as a result of the depolarization effect.

Wang et al.[104] included a periodical 180° domain pattern and depolarizing fields in their calculations and found that the size dependence of the Curie temperature and free energy is similar to that of single-domain ferroelectric films with positive extrapolation lengths. The Curie temperature drops to $0K$ below a critical film thickness.

2.2. SIZE EFFECTS IN FERROELECTRIC THIN FILMS AND STRUCTURES

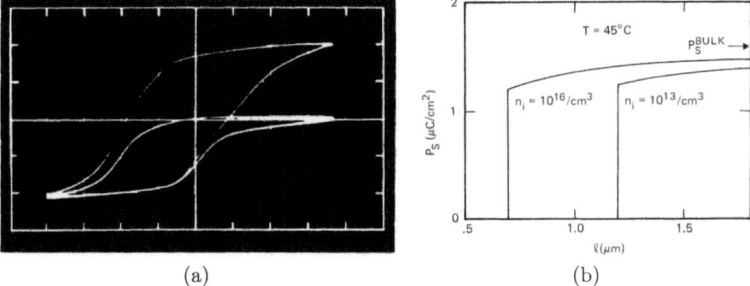

(a)   (b)

Figure 2.6: (a) Loop measured by Wurfel et al.[100] for $1\mu m$ TGS sandwiched between Au and p-type Si electrode. The influence of incomplete compensation is evidenced by acquiring the loop in the dark (small, asymmetric loop) and under illumination. (b) calculated size effect on the spontaneous polarization P_s due to incomplete charge compensation from the semiconductor at the interface for different carrier concentration n_i as a function of film thickness l.

2.2.2.2 Surface Effects

Depolarizing fields can not explain why sometimes ferroelectric stability can be even enhanced in some types of thin films at the surface.[96] This has been observed experimentally in KNO_3[96] where a ferroelectric phase was found in thin films but not in bulk samples of the same composition and has been treated theoretically using mean-field descriptions.[96,105] Tilley et al.[95] suggest that in some materials it appears that a surface layer orders before the bulk of the sample whilst in others, the opposite seems to occur. The surface of a material can not be treated as the bulk because of missing atomic neighbors. Unsaturated bonding states at the surface lead to its rearrangement which can inhibit ferroelectricity. Close to the surface, the translational symmetry is broken compared to the bulk. There, dipole-dipole interactions at the surface are different from the bulk and the polarization characteristics are changed.[80,94] For the last unit cell on the surface, the perpendicular polarization becomes more and more favorable when the thickness of the film is reduced.[72] Electron diffraction experiments on $BaTiO_3$ particles indicated that a surface layer of about $10nm$ having a much higher Curie temperature than the bulk affected the properties of small particles.[97] Space charge layers have been observed experimentally in single crystal $BaTiO_3$ by Triebwasser.[102] The $10-1000nm$ thick layers may result from high oxygen vacancy concentration at the surface.[103] Depolarizing fields may still be present in the case where free-surface charges are available to partially compensate the polarization discontinuity at the surface of the ferroelectric.[71] The local polarization near the surface is expected to occur over a distance comparable to the correlation length ξ of polarization fluctuations (a few lattice spacings well below the Curie temperature), see Figure 2.7, p.18.

The effect of the surface layer on the bulk is introduced in the Landau theory by an extrapolation length δ (see Figure 2.7, p.18). The extrapolation length measures the strength of the surface effect and describes the difference between the surface and the bulk. For the calculations, the extrapolation length is crucial. It is estimated to be between $1-10nm$ for perovskite

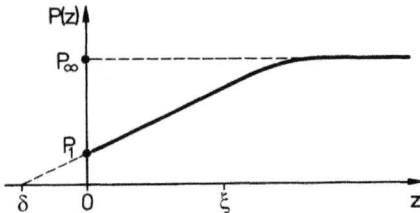

Figure 2.7: Variation of the local polarization $P(z)$ in the vicinity of a free surface situated at $z = 0$. The bulk polarization is denoted as P_∞ while P_1 denotes the polarization at the surface. δ is the so called extrapolation length, ξ the correlation length.[107]

oxides.[106] The extrapolation length is highly anisotropic because of the strong long range interaction in the polar direction and the weak one perpendicular to it.[76] Ghosez et al.[80] give an explanation of the extrapolation length from first principle calculations: negative short range interactions correspond to positive extrapolation lengths, positive short range interactions to negative extrapolation lengths. If short range interaction would be the only intersite interaction in a ferroelectric, it would have to be negative and only suppression of surface polarization could be observed. However, since the overall ferroelectric instability is governed by the combination of the short-range chemical and dipolar interactions, both, enhancement or suppression of the polarization can take place, depending on the material.
Negative extrapolation lengths may occur for sufficiently enhanced interaction strength at the surface layer. Then, the surface starts to order at a temperature higher than the bulk Curie temperature and the polarization at the surface may be enhanced.[100,105,107]
In the theoretical treatment, Kretschmer and Binder[107] and then Tilley and Zeks[82,105] first introduced the polarization behavior at the surface in the Landau theory, and Scott et al.[96] extended the second order ferroelectric-paraelectric phase transition to the first order by introducing a sixth-order term.
An insufficiency of many theoretical calculations[25,108] arises from ignoring the anisotropy of correlation forces. Li et al.[72,109] neglected depolarizing fields but included the highly anisotropic nature of the correlation forces in the Landau phenomenological theory, introducing an anisotropic extrapolation length. The critical size for both $PbTiO_3$ and $Pb(Ti_{0.5}Zr_{0.5})O_3$ were evaluated for infinite length with shrinking lateral size (long cell) and for infinite lateral length with shrinking thickness (thin film). The former case has a smaller critical size. This is illustrated for $Pb(Ti_{0.5}Zr_{0.5})O_3$ in Figure 2.8, p.19 . By letting $a_0 \to \infty$ the critical thickness of c_0 is obtained in a range of $10-15 nm$ at room temperature. For $c_0 \to \infty$, the critical value of a_0 is close to $2-3 nm$. For a given thickness, the Curie temperature decreases with decreasing a_0 and after a critical size the Curie temperature becomes $0K$. The Curie temperature is higher for greater thicknesses and equal lateral sizes. For thin film $PbTiO_3$, the critical lateral size at which the transition temperature is $0K$ was $4nm$ and for $Pb(Ti_{0.5}Zr_{0.5})O_3$ it was $8nm$. This figure illustrates the anisotropy of the extrapolation length: The lateral size effect (lateral dimension a_0) was cooperatively associated with the thickness of the cell c_0.
Huang et al.[110,111] studied the influence of surface bond contraction in $Pb(Zr,Ti)O_3$ for the whole compositional range. This phenomenon becomes important when the surface to volume ratio becomes very large. For a curved surface, the surface bond contraction will produce a strain along

2.2. SIZE EFFECTS IN FERROELECTRIC THIN FILMS AND STRUCTURES

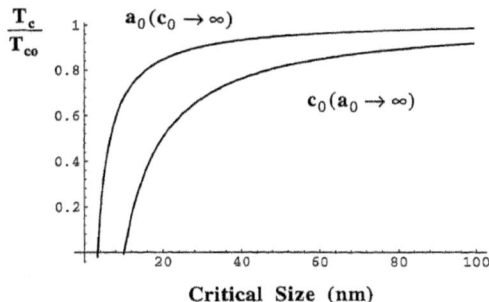

Figure 2.8: Calculated size effect in PbZr$_{0.5}$Ti$_{0.5}$O$_3$: Curie temperature dependence on film thickness c_0 (lateral dimension $a_0 \to \infty$) and slab lateral size a_0 (height $c_0 \to \infty$).[109] The slab has a smaller critical size than the film. This is due to the anisotropy of the extrapolation length.

the surface, creating a hydrostatic pressure on the inner part of the grain. Due to the surface tension, the pressure in the small particle varies in inverse proportion to the particle diameter. This pressure leads to an instable ferroelectric state.[112] Due to the surface bond contraction, the Curie temperature and the spontaneous polarization of tetragonal Pb(Zr,Ti)O$_3$ decreases with decreasing grain size for the whole compositional range and reached $0K$ for particles of $4nm$ for all Pb(Zr,Ti)O$_3$ compositions. This value is in accordance with the findings of Zhong et al.[106] ($4nm$) but disagreed with the findings of Ishikawa et al. ($12.6nm$)[113] and Chattopadhyay et al. ($7nm$).[114] Ti rich compositions have a higher Curie temperature than Zr rich ones. For small particles an increase of the Curie temperature is observed around the MPB. However, a fast drop is observed below a certain size. This is explained by the decrease of the tetragonality factor below a certain size and was observed also experimentally.[114] The dielectric constant in Ti rich tetragonal compositions showed an anomaly (a peak) at a small size. For 70 − 100% Ti composition this anomaly takes place at a particle size of $5.6nm$ indicating a phase change from ferroelectric (above $5.6nm$) to paraelectric. The phase transformation is observed by the reduction in tetragonal distortion c/a. Decreasing the grain size in the ferroelectric phase leads to an increase of the dielectric constant (dielectric anomaly). On the left side of the anomaly, the dielectric constant decreases with decreasing particle size. Experimentally, the contrary was seen for the situation above the anomaly:[114] the dielectric constant decreases with decreasing particle size. However, the calculation of Huang et al.[110] did not take into account domain wall contributions. Domain wall contributions can increase or decrease the dielectric constant. A high density of 90° domains increases the dielectric constant, but if pinning takes place, the mobility is greatly decreased and a decrease in the dielectric constant would be observed. The increase above $5.6nm$ is explained by the surface bond contraction which gradually leads to a phase change ferroelectric-paraelectric with the observed high permittivity. The measured decrease above the anomaly in the case of Chattopadhyay et al.[114] is explained by pinning effects of 90° domains which dominate in this size range.

Zhong et al.[106] and Rychetsky et al.[115] applied the Landau phenomenological theory to particles. Depolarizing fields were neglected. The extrapolation length δ is size dependent and becomes positive for small enough particles even if δ_∞ (extrapolation length for infinite size) is negative. The polarization is highest in the center and decreases towards the surface for $\delta > 0$.

If δ is negative, the polarization is enhanced towards the surface. With decreasing temperature, the polarization at the center approaches the bulk value. In the case $\delta < 0$, polarization may exist above the Curie temperature. The predicted theoretical values of the critical size were smaller (PbTiO$_3$: 4.2nm, BaTiO$_3$: 44nm) than the empirical fitting (PbTiO$_3$: 13.8nm, BaTiO$_3$: 115nm). This might be due to the fact, that this theory did not include effects from depolarizing fields.

Time resolved solutions of the Landau theory led to predictions with respect to switching time.[116] The Landau approach is used to investigate the situation where surface conditions can dominate over the bulk. The considered particle is flat with a lateral surface S in the $y - z$ plane. The border conditions in x are $\frac{dP}{dx} = \pm \frac{P}{\delta}$ with the extrapolation length[105,107] δ (see also Figure 2.7, p.18). For positive extrapolation lengths the polarization is reduced at the surface, for negative values, it is enhanced. Polarization enhancement near the surface results in different reversal times for the interior and surface. They found that more time is required for switching if the polarization is enhanced at the surface (negative extrapolation length). The minimum switching field required to switch both, the interior and the surface is smaller if the extrapolation length is large. Therefore, a large extrapolation length corresponds to a weak surface pinning. The contribution of the surface effects is inversely proportional to thickness, leading to a decrease in the minimum switching field, which approaches the bulk value for infinite thickness. They predict an observable double maxima in the polarization current during polarization reversal (switching): one for the interior (early peak) and one for the surface (later peak).

Ferroelectric finite sized cells in three dimensions are described by Wang et al.[25] with respect to the Curie temperature and the polarization. The polarization points in the c-axis direction. The depolarizing field is neglected since short circuited electrodes are considered. A single domain state is assumed. When $\delta > 0$, the observed transition temperature in small cells is smaller than the transition temperature of the bulk. In the case of $\delta < 0$, the polarization is enhanced at the film surface and the Curie temperature is higher than of the bulk. The size effect is the effect of both the thickness and lateral size. Only the lateral size effect is discussed and the extrapolation length $\delta \to \infty$. They perform calculations for second order ferroelectrics. The Curie temperature decreases with decreasing lateral dimensions and becomes zero at a finite value of the lateral size. The critical size, below which polarization is inhibited, is dependent on temperature. First and second order transitions yield the same dependence of the Curie temperature with respect to the lateral size. Near the cell boundary the polarization decreases quickly. For $a_0 = b_0$ and a second order ferroelectric at $0K$ the polarization drops to zero at a critical size in a continuous way, for a first order transition it jumps to zero at a critical size. The critical size is temperature dependent. Calculations of BaTiO$_3$ and PbTiO$_3$ showed a critical size at $300K$ of 9.1nm and 7.4nm respectively. If surface effects are to be included, the critical size is expected to be larger because of the positive extrapolation length in both materials. If cells with polarizations in the a-direction exist, the critical size is expected to increase with the number of such neighbors.

2.2.3 Experimental Observations of Size Effects

Size effects in ferroelectrics can be observed by determining the Curie temperature (with Raman scattering or heat capacity measurements) or the tetragonality (XRD). In the case of Raman measurements, a bulk sample is often prepared from small particles.[113,117] Several phonon modes are Raman active in the ferroelectric phase, but inactive in the paraelectric phase. Raman measurements are also suitable to determine the order of a transition (a first order transition would show an abrupt drop of the frequency shift at the Curie temperature).

2.2. SIZE EFFECTS IN FERROELECTRIC THIN FILMS AND STRUCTURES

Experimentally it is well known, that the polarization can be inhibited[113] or enhanced[96] in nanostructured materials. Tybell et al.[81] analyzed ultrathin sputtered, (001) oriented $PbZr_{0.2}Ti_{0.8}O_3$ films on conductive, Nb doped $SrTiO_3$. To circumvent the problem of leakage, the films were investigated with piezosensitve AFM. Ferroelectricity in a $4nm$ thick film was observed by poling a short line in the opposite direction. The opposite polarization was stable for at least $180h$. Chen et al.[87] investigated the nonlinear electric field dependence of piezoresponse in epitaxial $PbZr_xTi_{1-x}O_3$ ($x = 0.2, 0.4, 0.5$) on $SrTiO_3$ (100) and on $LSCO/SrTiO_3/Si$. No 90° domains existed in the films. The piezoelectric response decreased with an increase in electric field. d_{33} was measured by means of AFM and the nonlinearity increased with increasing Zr content. For Pb(Zr,Ti)O_3, Scott[118] reported that the MPB of Pb(Zr,Ti)O_3 is shifted from the bulk value 50% Ti to 20% for $100nm$ thin films. The transition temperature of KNO_3 films is higher than that of the bulk, and some samples exhibit an increase in T_0 with increasing film thickness and others a decrease.[96] δ measures how far the "extra" surface polarization extends into the film. Roelofs et al.[59] found experimentally that a $PbTiO_3$ grain on Pt (111) of about $18nm$ and $5nm$ in height didn't show any piezoelectric response and attributed this to be the critical size. $PbTiO_3$ particle sizes as small as $14nm$ in average size could be produced.[117] Raman measurements are suitable to observe the Curie temperature. In bulk $PbTiO_3$ samples, the Curie temperature was $T_0 = 493°C$. The $14nm$ small $PbTiO_3$ particles showed a Curie temperature considerably lower ($T_0 = 436°C$). The critical size of complete inhibition was extrapolated to $10.7nm$. XRD measurements can be used to measure the decrease in the tetragonality c/a upon downsizing. The tetragonality measured by XRD decreased at small particle sizes and extrapolation showed the expected inhibition of the ferroelectric phase for $PbTiO_3$ particle size of $11.7nm$. The grain size affects the switching properties of thin films[119] and bulk ceramics.[120] For $PbTiO_3$ thin films (not freestanding), random orientations were observed.[121] For very thin films on NaCl (grain size of about $80nm$), the (200) $PbTiO_3$ peak disappeared, and c-domains were observed in $\theta - 2\theta$ XRD diffraction measurements. The film was still tetragonal but no a-domains existed. The same was observed on $Pt/SiO_2/Si$ but here the $PbTiO_3$ was (100) oriented.

Size effects were also observed in $Ba_xSr_{1-x}TiO_3$.[122] The transition temperature decreased with grain size and the transition becomes diffuse. The boundary condition at the interface ferroelectric/electrode plays a decisive role in the manifestation of size effects.[123] For the case of normal metallic electrodes, the polarization vector at the interface is $P = 0$. This corresponds to a strong edge field resulting in the freezing out of the ferroelectric polarization at the interface and thus exhibiting severe size effects. However, for conductive oxide electrodes, the polarization would not vanish at the interface and therefore the size effects are largely suppressed.[123] This was shown in $SrTiO_3$ sandwiched between Pt and $YBa_2Cu_3O_{7-x}$ (YBCO). The sandwich structures were grown on single crystalline $SrTiO_3$ (100). They showed experimentally that the size effect (film thickness) on the dielectric constant was reduced when using oxide electrodes, i.e. the decrease of the dielectric constant of $SrTiO_3$ decreased at thinner film thickness than in the case of metallic Pt electrodes.[123] A lateral size effect in switching time was found.[124] The theory predicts a time decrease with decreasing size. Considering that switching is done by domain wall motion, switching would also be subjected to a size effect. An exponential relationship between switching time and applied field was observed In $BaTiO_3$,[125] where an activation field parameter in the exponent can be defined. The necessary activation field necessary was found to increase in a hyperbolic manner as the film thickness decreases.

The grain size of nano-sized powders of $Pb_{0.8}Ba_{0.2}TiO_3$, $Pb_{0.8}Sr_{0.2}TiO_3$, $Pb_{0.8}La_{0.2}TiO_3$ and PZT was studied by means of Raman spectroscpoy.[126] In all cases, the Curie temperature decreased when the particle size decreased below $60nm$. Zhong et al.[127] fabricated ultrafine $PbTiO_3$ particles. The tetragonality at room temperature decreased with decreasing particle

size. The critical size at which $T_0 = 0K$ was extrapolated from specific heat measurements and was found to be $9.1nm$.
A size effect at an extremely large scale was found in the II-IV phase transition in $CsZnPO_4$.[128] The phase transition, detected with heat capacity measurements, was inhibited single crystals below $0.1mm$.
The decrease of the dielectric constant with film thickness was found in $Bi_2VO_{5.5}$, $Bi_4Ti_3O_{12}$ and Bi_3TiNbO_9 epitaxially grown on $SrTiO_3$ (100).[129]
Size effects of $BaTiO_3$ thin films is determined by the size of the crystallites.[130] Varying the substrate temperature during sputtering changed the crystallite size (low T low size). The dielectric constant was found to decrease with decreasing crystallite size showing a critical value of $30nm$.
Size effects have been observed in $SrBi_2Ta_2O_9$ (SBT) particles.[131] The transition temperature of SBT particles markedly decreased as the particle size decreased. The qualitative understanding is found by Zhong et al.[127] From an empirical expression, the particle critical diameter where the transition temperature drops to $0K$ was determined empirically at $2.6nm$. This is much less than in $PbTiO_3$ ($13.8nm$[113]) and even smaller than in $Pb(Zr,Ti)O_3$[106,110] ($4nm$).
Jiang et al.[132] investigated small $PbTiO_3$ particles by TEM. Particles with diameters of about $15nm$ were still tetragonal ($c/a = 1.052$), at $30nm$ it was $c/a = 1.056$. No amorphous surface layers were found on the particles. 90° domain walls were observed in particles down to $20nm$. No 90° domains were observed below that size. The domain size decreases with particle size. Only two domains were found in a $26nm$ large particles, whereas a $200nm$ large particle contained three domains. The domain wall width was found as $14Å$. The tetragonality c/a decreased from the bulk value 1.0652 to 1.052 for particles of $15nm$ and $30nm$.
Size effects have been observed in $BaTiO_3$ ceramics.[26] The dielectric constant was increased considerably when the grain size approached $1\mu m$ (increasing density of 90° domains and stress effects acting on the domain wall mobility[26,133,134]) and was decreased (vanishing of 90° domains) for further decrease in grain diameter. The smaller the grains the more the dielectric and the elastic constants are determined by the contribution of 90° domain walls. The domain density is increased for smaller dimensions D[135] and the domain size is decreased $\propto \sqrt{D}$. Experimentaly, in $BaTiO_3$ the domain width decreases for small particles and approaches a constant value for big particles.[26] For particles smaller than $1\mu m$ the tetragonality decreases. Extrapolating the data suggests that no 90°-domains exist anymore below particle sizes of about $400nm$. On the one hand 90° domain walls contribute to the dielectric constant,[26] so an increase of a-domains would increase the dielectric constant. On the other hand, these domains have to be mobile to contribute to the dielectric constant. High stresses in fine grained ceramics create a high density of 90° pinned immobile domains. The high stress can suppress the spontaneous polarization and forcing the grains back towards a cubic state. This corresponds to a vanishing of a-domains and to a size induced phase transformation ferroelectric-paraelectric where high dielectric constants are measured.[26,133] However, Arlt et al.[26] explain the increase to mobile 90° domains.
Ferroelectric $Pb_5Ge_3O_{11}$ has been prepared by heating a quenched glass state to $510°C$. Temperature controlled rapid nucleation and slow crystal growth led to very fine grained ceramics with grain sizes down to $10nm$. In large grained ceramics, normal ferroelectric behavior was found whereas in fine grained ceramics a very diffuse phase transition and polarization relaxation were observed.[75]

2.3 Nucleation and Growth of Sputtered Epitaxial Thin Films

Thin films are formed on a substrate by a process of nucleation and growth.[136,137] Nucleation is theoretically expressed in the frame of the classical nucleation theory. Epitaxial growth is the deposition of a film on a single crystalline substrate where the atomic arrangement of the film is given by the few topmost layers of the substrate. One distinguishes homoepitaxy from heteroepitaxy. In the former case, film and substrate are the same material, in the latter case they are different. In this work we deal with heteroepitaxial growth, named epitaxial growth for simplicity. Homoepitaxial growth of ferroelectric TGS was observed as early as 1977 on metal layers having pin holes down to the single crystalline TGS substrate. When spinning the TGS solution, crystallization started in the pinholes at the contact with TGS and the grains grow up the pinhole and then laterally on the metal.[138,139] In the initial stage of thin film formation the formation of small clusters of the film material from individual atoms or molecules takes place. As time progresses, more clusters are nucleated and these clusters grow, coalesce, and finally a continuous film is formed which then thickens. Phenomenologically, three growth modes can be distinguished: Van-der-Merwe, Volmer-Weber and Stranski-Krastanow growth modes. In the case of Van-der-Merwe growth, the interaction between the substrate's atoms and the ones of the film is stronger than between the atoms of the film, and the film material has a high wetting coefficient. As a consequence, layer by layer growth takes place, also called 2D-growth. This is in contrast to the Volmer-Weber growth, also called island growth. During a Stranski-Krastanow growth, initial 2D growth leads to the formation of a wetting layer. After a critical thickness layer by layer growth is not possible any more due to large discrepancies of the lattices. The high stress is relaxed through dislocation formation and island growth. Stranski-Krastanov growth of Ge on Si led to 3D growth after $4ML$. Triangular shaped (self assembled) Ge islands can be obtained in that way.[140] Theoretical models of epitaxial growth suggest that the growth model is determined by the free energy of the substrate surface (ρ_s), the interfacial free energy (ρ_i), and the surface free energy of the heteroepitaxial layer (ρ_f). The inequality

$$\rho_s > \rho_f + \rho_i$$

sets the condition for the epitaxial film to wet the substrate. In this case, Van-der-Merwe growth may occur. If the sign is opposite, one usually obtains Volmer-Weber growth, *i.e.* no wetting of the substrate takes place. The Stranski-Krastanov growth generally occurs when there is wetting of the substrate but the overlayer strain unfavorable.

2.3.1 General Model

In this work we discuss systems in the frame of the classical nucleation theory in which the film is created via a constant flux R of atoms or molecules arriving on the substrate surface. Atoms accommodated on the surface are called adatoms. The population of single adatoms or molecules n_1 (cm^{-2}) builds up and they diffuse over the substrate at temperature T, with the diffusion coefficient D (cm^2/s). Re-evaporation of the accommodated species occurs, and the mean residence time is τ_a. Random collisions between adsorbed species form small clusters of a population n_j (cm^{-2}, j: number of atoms) form. Two competing mechanisms determine the stability of a nucleus: on one hand, a the system needs energy to create the nucleus' surface (which is not favorable for cluster formation), on the other hand, binding energy is gained by the system. The total free energy of a spherical nucleus of radius r can be written as

$$\Delta G_0 = 4\pi r^2 \rho_{cv} + \frac{4}{3}\pi r^3 \Delta G_v \quad \text{with} \quad \Delta G_v = -\frac{kT}{V}\ln\left(\frac{p}{p_e}\right) \tag{2.1}$$

where ρ_{cv} is the surface energy between the nucleus and the vapor, r the radius of the nucleus, ΔG_v is the free energy gained by binding, p/p_e is the supersaturation, where p is the vapor pressure, and p_e the equilibrium pressure, and V is the atomic volume.

The surface energy of small clusters increases by the square of its radius and decreases with the volume energy (binding energy E_j) by the cube. For small clusters it is favorable to decrease the total energy by dissociation, however after having reached a critical size growth becomes more probable than decay. Consider a perfect substrate containing N_0 (cm^{-2}) site density of equal adsorption (or desorption) sites with adsorption activation energy E_a. The mean residence time τ_a of a molecule or adatom is given by:

$$\tau_a = \nu_0^{-1} \exp\left(\frac{E_a}{kT}\right)$$

τ_a: mean residence time
ν_0: rate constant for desorption
E_a: activation energy of ad- or desorption
k: Boltzmann constant $1.38 \times 10^{-23} J/K$
T: Substrate surface temperature

Now consider a single adatom on a surface which could jump to ξ energetically equivalent adsorption sites which are located at a distance a_0. The mean time between hops can be written as follows:

$$\tau_d = \nu_1^{-1} \exp\left(\frac{E_d}{kT}\right)$$
$$\nu_1 = \xi \, \nu_d$$

τ_d: mean time between hops
ν_1: rate constant for hops in all directions
ν_d: rate constant for hops in any one direction
ξ : number of possible hopping directions
E_d: activation energy for diffusion

To relate the mean time between hops τ_d to the diffusion D we consider Fick's law:

$$\mathbf{j} = -D \, \mathbf{grad} n_1$$

$\mathbf{j}$: flux density directed to the downward gradient
n_1: adatom concentration
D: diffusion coefficient

It is found that the adatom diffusion on a high symmetry crystalline substrate is characterized by a single value of D, i.e. it is isotropic.[141]

$$D = \frac{1}{4} a_0^2 \tau_d^{-1} = \frac{1}{4} a_0^2 \nu_1 \exp\left(-\frac{E_d}{kT}\right) = \frac{1}{4} a_0^2 \xi \, \nu_d \exp\left(-\frac{E_d}{kT}\right)$$

The factor $\frac{1}{4}$ is dependent of the growth dimension, here for 2D-growth. For 1D this factor is $\frac{1}{2}$ and for 3D it is $\frac{1}{6}$. The jump distance a_0 is dependent of the crystalline structure of the surface. In the case of a Pt (100) surface, a_0 is equal the de lattice spacing d_{100}. For a Pt (111) surface,

2.3. NUCLEATION AND GROWTH OF SPUTTERED EPITAXIAL THIN FILMS

this length is $\frac{1}{\sqrt{2}}d_{100}$.

For surface diffusion issues we are interested to know the mean diffusion length before desorption. Assuming the atomic movements on the surface as a 2D gaz, the mean diffusion path can be estimated by Einstein's relation about brownien movements $\sqrt{\langle x^2 \rangle} = \sqrt{2D\tau_a}$, and we find:

$$\sqrt{\langle x^2 \rangle} = \frac{a_0}{\sqrt{2}} \frac{\nu_1}{\nu_0} exp\left(\frac{E_a - E_d}{2kT}\right)$$

Considering the sputtering of Pb(Zr,Ti)O$_3$ the key species are the PbO molecules which determine the nucleation and growth. This is a deviation of above hypothesis of deposition of equal species. Nevertheless, we can estimate the mean diffusion distance of a PbO species on a Pt (111) or Pt (100) surface. As a first estimation, the desorption rate ν_0 and the rate constant for hops in all directions is often supposed to be equal.[141] It is further estimated that $E_d = E_a/4 = 0.45eV = 7.2 \times 10^{-20} J$. The hopping distance on the (100) plane is $a_{0,(100)} = 3.92$Å, and on the (111) plane $a_{0,(111)} = \frac{1}{\sqrt{2}} a_{0,(100)} = 2.77$Å. Then, we find a mean diffusion distance for a substrate at 570°C

$$\sqrt{\langle x^2 \rangle_{(100)}} = 3.1 \mu m$$

$$\sqrt{\langle x^2 \rangle_{(111)}} = 2.2 \mu m$$

2.3.2 Epitaxy on Oxyde Ceramics: A Phenomenological Approach

The calculations made before were based on a very general model:

- a perfect atomic surface with adsorption site density given by the surface
- the binding energy within the cluster

These assumptions are far away from a model for epitaxy. It could also be applied for an amorphous surface with a mean hopping distance be defined. It can not predict whether epitaxial growth is possible or not because it does not include

- lattice mismatch
- chemical bonding
- crystalline periodicity of the surface, the effect of epitaxy
- stress, and relaxation through dislocation formation
- surface energy of the substrate
- surface energies of the grown material's crystal planes
- interface energies

In what follows, the special situation for epitaxial depositions on oxide ceramics is highlighted. A straightforward factor in predicting epitaxial depositions is lattice mismatch. But good lattice match alone is not the only criterion for epitaxy, especially for depositions on oxide ceramic. For example, it was shown, that Pt (100) was difficult to grow epitaxially on $SrTiO_3$ and MgO (100) substrates despite good lattice match and similar crystal structures. Here, the decisive factor is the low surface energy of the Pt (111) which inhibits (100) growth if the deposition conditions are not correctly adjusted. Epitaxial growth has been observed for large lattice mismatch due to the high oxygen affinity of the metal with the oxide substrate. The surface energy of solids is defined as the excess energy of a free surface compared to that of the bulk. The knowledge of this quantity is of great importance as it controls many phenomena such as the equilibrium form of the crystal, cleavage, crystal growth, epitaxy, surface diffusion, wetting, reactivity of solids, adsorption and catalysis.[142] Atoms near the surface of a crystal are under the influence of other forces than those in the bulk, and in most cases relaxation and reconstructions of the bulk lattice will occur in the topmost layers.[143] Surface energies have been calculated for fcc metals like Cu, Ag, Ni, Pd and Pt.[143] In all cases the close packed (111) face has the lowest surface energy, followed by the (100) and the (110) faces. The calculated value are in agreement with the "average" solid-vapor surface energy derived from the data of the liquid surface tension of metals at the melting point.[144] The surface energy is written as a sum of two parts:

$$E_s = E_c + E_r$$

E_c (positive) expresses the energy necessary for cutting the bonds perpendicular or inclined to the surface without moving the atoms with respect to their neighbors in the separated parts of a crystal. E_r (negative) is the energy gain due to relaxation.[142] E_r is dominant if a metal with a great oxygen affinity is deposited on an oxide ceramic. If we are interested in the equilibrium form of a crystal, we would adress E_s. To analyze the reactivity of a surface, E_r is the parameter to be considered; it is a measure of unexploited binding ability.[142] Due to the good fitting of the condensate's lattice constant with the substrate the interface energy can be lowered and nucleus with a size as small as 1 atom can be stable.[145] The interface[146] between the formed film and the substrate becomes a crucial issue in epitaxial film formation. There are two main issues which govern the system:

- Lattice mismatch
- Chemical bonding

2.3.2.1 Coherent, Semicoherent and Incoherent Interfaces

The interface between an epitaxial thin film on a substrate can be coherent, semicoherent or incoherent depending on the lattice mismatch between corresponding planes of both constituents and the bonding strength at the interface.[147,148] The lattice mismatch can be defined as

$$\Delta m = \frac{a_f - a_s}{a_s}$$

with a_s, a_f being the lattice constants of the substrate and the film. The lattices of the constituents are in general incommensurable, that is the lattice constant of the ceramic substrate is not an integer multiple of the lattice constant of the metal. In the case of coherent interface, the lattice mismatch between a metallic film and a ceramic substrate is accommodated entirely

2.3. NUCLEATION AND GROWTH OF SPUTTERED EPITAXIAL THIN FILMS

by straining the lattice of the film.

Coherent interfaces possess periodicity parallel to the interface. Semicoherent interfaces are characterized by coherent regions separated by misfit dislocations as observed for Ag on MgO (100)[149] for $GaAs_{0.5}P_{0.5}$ on GaAs.[150] The lattice in the bulk of the film is not strained if the misfit dislocations accommodate the entire lattice mismatch. In this case arrays of misfit dislocations have be observed experimentally for Nb on Al_2O_3 interfaces.[151] Incoherent interfaces can be thought of as an interface between two rigidly connected lattices. Incoherent interfaces can also be interpreted as an extreme case of a semicoherent interface in which the displacement field of the misfit dislocation vanishes, that is the dislocation core of the misfit dislocation is completely delocalized.[148] The state of interface depends, first of all, on the physical and chemical nature of the contacting phases (including size and symmetry of elementary crystal cells), but also on such external factors as temperature or even the thickness of the contacting crystals. In the last case, there exists a critical thickness (for epitaxial crystal films growing on thick substrates) above which the presence of misfit dislocations at the interface becomes energetically favorable.[147,152,153]

2.3.2.2 Dislocation Formation

The formation of misfit dislocations is a process that occurs as a consequence of the discrete nature of the crystalline material[153] to relax misfit strains. For a film thickness h greater than some critical value h_c the effect of the elastic mismatch stress dominates the effect of the self-stress (the equilibrium stress field around dislocation), and the threading segment tends to advance along the glide plane. For $h < h_c$ the reverse is true and the threading segment tends to recede. The density of misfit dislocations at the interface decreases with with the size of the mesa.[153] Misfit across the interface between an epitaxial film and its substrate is accommodated by uniform elastic strain until a critical film thickness is reached.[146,150,154,155] Thereafter, it is energetically favorable for misfit to be shared between dislocations and strain. Dislocations at the interface are not necessarily deleterious, in contrast to threading dislocations which extend across of a thin film. Such dislocations can act as paths of relatively easy charge transport, as diffusion paths for dopants or as surface seeds for defects in subsequent overgrowth. Chances of complete coherency across an interface are best for lattice-matched systems, which have materials with the same crystal structure and nearly identical lattice parameters. However, in heterostructures, a small mismatch of lattice parameter between a substrate and a film being deposited on it could be accommodated by elastic strain of the film as it grows. On the other hand, stress induced in the material by this mismatch elastic strain can serve as a driving force for nucleation and growth of crystal dislocations.

2.3.2.3 Metal/Ceramic Interfaces

The binding of a ceramic to a transition metal is less well understood, since here, strong covalent pd-bonds may be formed across the interface between oxygen and the transition metal.[156] To understand what happens at metal-ceramic interfaces, first-principles calculations of the electronic charge density, the total energy and the relaxed structure[157] has been done for Ag, Ti on MgO (100)[156] and Nb on Al_2O_3.[157] Calculations for metal-ceramic interfaces are time consuming and possible only for a limited number of model systems[149,158] where for instance thermal stresses can be neglected[148] such as AB-initio calculations for Al or Mg on $MgAl_2O_4$[159,160] and for Ti

28 CHAPTER 2. THEORETICAL BACKGROUND AND STATE OF THE ART

or Ag and Cu on MgO (100)[156] or MgO (222) polar metastable surfaces (using first principles local-density functional theory),[161,162] metals on AlN,[163] Nb on Al_2O_3[151,164] . The lattice misfit is only 2% between fcc Ti ($a = 4.21$Å) and MgO (100) and 3% between fcc Ag ($a = 4.21$Å).[156] Misfit dislocations at interfaces between weakly bonded partners are likely to be delocalized. If there were no adhesion across the interfaces the interface energy would equal the sum of the surface energies. However, the interface energy was in the same range as the individual surface energies. This means that binding happens as well between Ti/MgO and Ag/MgO. Titanium was found to bind more strongly than Ag. Both Ti and Ag bond on top of O. Binding between Ti and O is predominantly covalent and weaker than in bulk titania. The bonding between Ag and O is weak and predominantly ionic.[156]

The oxygen affinity may be used to predict trends for interfacial energies and epitaxial orientations.[149,158] It is defined as the dissociation pressure of the oxide at $1000K$ for equilibrium between a metal and its oxide. For a given dissociation pressure a particular epitaxial orientation may be favored. As long as the ratio of lattice mismatch and oxygen affinity of the metal is below a critical value, epitaxial growth can be observed. For metals with very high oxygen affinities, other mechanisms, such as strong interdiffusion or the formation of interfacial phases may dominate the growth process.[158]

2.3.2.4 Nb on Al_2O_3 and Nb, Cu on TiO_2

This system is widely discussed in the literature and is shortly reviewed in this work. Nb on different Al_2O_3 surfaces serves as a model system since the thermal expansion coefficient of both materials are nearly identical.[165] All Nb-α-Al_2O_3 interfaces were semicoherent with dislocation networks accommodating the misfit between Nb and α-Al_2O_3. The misfit strain ranged from 2 to 12%.

In the case of Nb on TiO_2 (110),[165] where a large lattice mismatch exists, a high defect density was found at the interface but Nb still grew epitaxially on TiO_2. Moreover, Nb grew in the two dimensional mode in the first two monolayers. This can be explained with the strong chemical interaction between Nb and TiO_2 due to the strong oxygen affinity of Nb.[157] One important characteristic that makes metal interfaces with the TiO_2 surface of particular interest is the fact that Ti^{4+} can be reduced by a large number of metals. There exist a reaction layer with a high density of defects of $2nm$ of reduced TiO_2 and $1nm$ thick layer of oxidized Nb. For Cu depositions on TiO_2 (110) no such reaction interface is formed[165] due to the low oxygen affinity of Cu. Three dimensional growth of Cu happens. An epitaxial incoherent Cu/TiO_2 (110) interface without dislocations is created in this case. Metals that are highly reactive with oxygen show a strong oxidation/reduction reaction at the interface and therefore a high wetting ability is expected.[165] Low wetting ability and agglomeration in three-dimensional islands are expected for less reactive metals having weak interaction with the semiconducting substrate.

For Nb on Al_2O_3 (0001) is was found[157] that for O termination of the sapphire surface, the modification to the electronic and atomic structure of Nb is several layers deep and the sapphire is not much perturbed below the top layer. In the case of an Al termination, the perturbation of the Nb lattice and its electronic structure is much weaker.

2.3.2.5 Epitaxial Pt on Single Crystalline Substrates

This system is also used in this work (see the additive route in Figure Figure 4.16, p.94) For the orientation of the thin film the crystal plane with the lowest surface energy is the most stable

2.3. NUCLEATION AND GROWTH OF SPUTTERED EPITAXIAL THIN FILMS

one, and will be the fastest growing. Atomic high density planes are more stable than planes with lower density due to the saturation of bonding. In the case of fcc metals like Cu, Ag, Au, Ni, Pd, Pt and Al, surface energies calculated from first principle calculations showed that the (111) planes are most stable. This is in accordance to experimental observations of the oriented (111) growth of Pt on amorphous SiO_2 where interface energies are not reduced by chemical or epitaxial effects, and in our experiments for Pt (111) on $SrTiO_3$ (111). Surface energies for (111) Au, Pt and Ir have been calculated:[166] Au ($0.078eV/Å^2$), Pt ($0.137eV/Å^2$), Ir ($0.204eV/Å^2$). Using local density calculations the Pt (100) surface energy has been calculated[167] and found to be $0.147eV/Å^2$ ($1.126eV/atom$). Other work[168] find values between 0.110 and $0.158eV/Å^2$. It is therefore more difficult to obtain (100) orientation of Pt.

Pt is an interesting material not only as bottom electrode for perovskite materials but also as a model catalyst.[169] Pt was grown epitaxially on $SrTiO_3$ (100) at high temperatures.[123] At low temperatures, mixed (111) and (100) growth occured[170] by RF sputtering. Pt has a small lattice mismatch with $SrTiO_3$ of 0.46% (lattice parameter $Pt = 3.905Å SrTiO_3 = 3.923Å$), however, Pt has a weak oxygen affinity. The same lattice mismatch exist between Pt and SrO or TiO_2 terminated surfaces. Different growth behavior on SrO or TiO_2 terminated $SrTiO_3$ can be interpreted in terms of a change in interfacial energy due to a change in the degree of chemical bonding across the interface. The surface energies of Pt (111), (100) and (110) have been calculated to be $1.191J/m^2$, $1.476J/m^2$ and $1.651J/m^2$ with a difference of $0.285J/m^2$ between (100) and (111). On TiO_2 terminated surfaces (100) Pt was grown, on $2ML$ SrO terminated surfaces the Pt was mixed (100) and (111) oriented. Without considering interface energies and chemical bonding, only (111) growth would be expected. The good lattice match of Pt (100) with the substrate creates a low interface energy with the TiO_2 terminated surface. A high bond strength exists for the (100) orientation and TiO_2 surfaces. On SrO the bond strength is lower and the interface energy is almost the sum of the individual surface energies, which makes (111) Pt orientation more probable.[171]

Epitaxial Pt was also grown on (100) rocksalt (NaCl) covered with Ag by electron beam evaporation.[172] Pt single crystals with a thickness of up to $20\mu m$ could be grown in that way at substrate temperatures of above $200°C$. Pt was evaporated directly on cubic, octahedral and dodecahedral planes of rock salt[173] at $6nm/min$. Island growth was observed at all temperatures and epitaxy was obtained above $430°C$. Recently, epitaxial Pt (100) was deposited by RF sputtering on Si using a $\gamma-Al_2O_3$ buffer layer.[42] Low growth rates and high temperature favored the (100) growth. (100) orientation was predominant for films grown by electron beam evaporation at a rate of $0.6nm/min$. above $400°C$[169,174] and islands with cubic shapes were observed. Pt was grown epitaxially on LiF and MgO (100) by electron beam evaporation at $450°C$ and a growth rate of $6nm/min$.[175] Pt was deposited by PLD[176] on MgO (100). Epitaxial growth was obtained at $400°C$ for $0.03mTorr$. Above and below, the (111) orientation appeared (too high or too low surface mobility). Low pressures were required to generate epitaxial (100) growth. At low temperatures and higher oxygen pressures platinum oxides were created. Platinum was found to grow in the Volmer-Weber mode on MgO (100).[177] Two epitaxial relationships are generally observed on MgO (100): cube on cube (100) and (111) with $[110](111)Pt\|[110](100)$ MgO relationship. The (100) orientation tends to appear at low deposition rates and high temperatures. The two orientations coexist over a wide range of substrate temperature of $500-700°C$ by electron beam evaporation. This is thought to be because of a competition between interfacial and surface energies. Depositions were carried out by electron beam evaporation at $700°C$ and $12nm/min$. $4nm$ thick Pt on MgO (100) grew epitaxially, for thicker films ($10nm$), (111) orientation also appeared. Calculations showed that cube on cube (100) orientation has a lower interface energy with MgO than (111). Pt forms spherical caplets on MgO (100), where no atomic place defines the surface. However, when the film coalesces from islands, the roughness decreased and the surface energy becomes more important. Surface energy anisotropy may then influence the system, and as in fcc crystals the (111) plane is the lowest energy plane, the film will switch to (111) growth. Ahn et al.[178] investigated sputtering of Pt on MgO (100) films as a function of deposition rates and temperature at an Ar pressure of $12mTorr$. The thickness of the films were around $250nm$. At high temperatures ($800°C$), (100) films were obtained where a switching to (111) orientation was observed at $400°C$. The same in plane orientation were found than by McIntyre et al.[177] Lower deposition rates favor Pt (100) growth.[178] For the deposition of (100) oriented $Bi_4Ti_3O_{12}$, Pt and Ir were epitaxially grown on epitaxial $(ZrO_2)_{1-x}(Y_2O_3)_x$ (YSZ) on Si (100)[179] by sputtering in an Ar atmosphere at $700°C$ (Pt) and at $600°C$. Films on Ir didn't show a ferroelectric loop, on Pt they did. This is probably due to an oxidation of the Ir surface. Cube-on-cube growth of Ir on MgO (100) was obtained by an RF sputtering process.[180] Below $400°C$, mixed (111) and (100) Ir was observed, above that temperature, (100) growth was observed with few heteroepitaxial grains oriented $(221)[\bar{1}\bar{1}2]Ir\|(100)[\bar{1}10]MgO$. The XRD FWHM was $0.5°$ and the measured (100) lattice parameters were $3.839Å$ (Ir) and $4.211Å$ (MgO). The (100) lattice parameter measured was $3.839Å$ (Ir) and $4.211Å$ (MgO). Giant Ir crystals were obtained at $500°C$ showing lowest energy (111) facets. The cubic base of the pyramidal Ir crystals were rotated by $45°$. $Pb(Zr,Ti)O_3$ on MgO (100) covered by epitaxial Pt (100) was deposited by RF-magnetron sputtering.[181]

2.4 Pb(Zr,Ti)O$_3$ on Single Crystalline Substrates

The deposition of epitaxial $Pb(Zr,Ti)O_3$ and $PbTiO_3$ on single crystalline substrates has been extensively investigated in the past. Deposition methods like sol gel,[182–188] various CVD methods,[87,189–197] PLD,[87,187,198–201] sputtering from a single $Pb(Zr,Ti)O_3$ target[181,202–204] and from multiple metallic targets[205] are reported. $Pb(Zr,Ti)O_3$ was deposited on various (001) oriented single crystalline substrates,[206] such as MgO, $SrTiO_3$,[207] $LaAlO_3$, $LaNiO_3$, sapphire. On Si, lead diffusion into Si makes it impossible to deposit high quality $Pb(Zr,Ti)O_3$, and barrier layers are necessary. Recently epitaxial depositions on Si with numerous intermediate buffer layers been fabricated.[185,192,195,208]

Foster et al.[183] investigated sol-gel $Pb(Zr,Ti)O_3$ in the whole composition range on $SrRuO_3$ buffered single crystalline $SrTiO_3$ (001). c-axis oriented films were obtained and for tetragonal compositions the same surface pattern was found as in our work (see Figure 3.6(e), p.49). All films showed a-domains and their amount was estimated to 17%. A low coercive field of $50kV/cm$ was measured on a film with 65% Ti (see Figure 3.13, p.55) and a low remanent polarization of $55\mu C/cm^2$. The a-lattice parameter was slightly higher than the bulk value (see Figure 2.9, p.31). The most extensively investigated substrates, which are relevant for this work, are $SrTiO_3$ (100) and MgO (100). On both substrates, epitaxial depositions of all $Pb(Zr,Ti)O_3$ compositions have been demonstrated.

2.4.1 Domain Formation

The domain structure in ferroelectrics is essential for the electrical and mechanical properties. Some ferroelectrics such as PZT allow the formation of so called $90°$ domains where the polarization vectors of adjacent domains form an angle of $90°$. This is possible due to a small tetragonal deformation, i.e. when the a-axis is only slightly smaller than the c-axis. This is the case in PZT. The corresponding domain walls contribute greatly to the dielectric constant.[210] A laminar domain structure is usually seen in very small grains.[211] The experimentally observed

2.4. PB(ZR,TI)O$_3$ ON SINGLE CRYSTALLINE SUBSTRATES

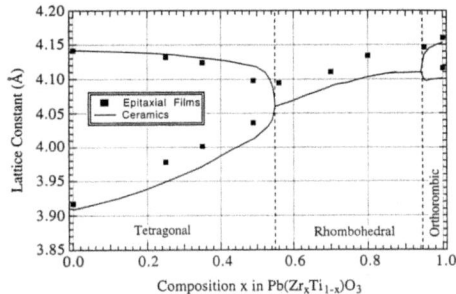

Figure 2.9: Lattice parameters of Pb(Zr,Ti)O$_3$ for thin films and bulk over the whole compositional range.[209]

decrease in the dielectric constant can be due to inhibited domain wall motion (pinning) when the grain size approaches the width of the domain wall. A competing mechanism, which tends to increase the dielectric constant at small sizes, is the surface bond contraction effect.[111] Several models have been proposed, including the presence of internal stresses in fine-grained ceramics, which are due to the absence of 90° domains walls,[133,134] the domain-wall contribution to the dielectric response in fine-grained ceramics[26,211] and shifts of the phase transition temperatures with grain size. Domain formation, especially 90° domain formation in epitaxial films undergoing a phase transformation is a mechanism that relaxes the total strain energy,[212] which is a result of the phase transformation, lattice mismatch, and the difference in the thermal expansion coefficients of the film and the substrate. Strain can be released by domain configurations, domain tilting and interface dislocations. The concept of domain formation in constrained layers was first developed by Roytburd et al.[213] and later applied to epitaxial ferroelectric films.[214,215] Stress strongly affects polydomain formation and can suppress it completely, allowing one to obtain single domain structures.[216] Table 2.1, p.32 gives an overview of the crystallographic and thermal expansion of the materials used in this work.

The lattice constant of the substrate and the difference in the thermal expansion coefficient play a crucial role in domain formation. If the thermal expansion coefficient of the substrate is larger than that of the film, as in the case for PZT on MgO and SrTiO$_3$ substrates, it is expected that horizontal compressive mechanical stress along the plane of the film favors the accommodation of the shorter a-axis in the plane during cooling through the phase transition temperature. This leads to c-axis oriented films. Tensile stresses (for example for depositions on Si) favors the c-axis to be in the substrate plane.[217] The development of the domain structure in 300nm thick PbTiO$_3$ on MgO (100) as a function of temperature was shown using synchrotron measurements by Lee et al.[192] At a temperature above the transition temperature the cubic phase was observed. The large misfit of about 0.2Å induces a high tensile stress which is partially reduced by dislocation formation. The relaxation of the film by dislocation formation can be expressed in terms of a effective lattice parameter of the substrate. The introduction of misfit dislocation partially accommodates the lattice misfit and reduces the coherency strain energy at the growth temperature. Following a model by Matthews,[150] the equilibrium dislocation density can be calculated. The dislocation density that would completely relieve the entire misfit in one dimension at the growth temperature would be $1.6 \times 10^6 cm^{-1}$ in the PbTiO$_3$ MgO system at

		Lattice Pb(Zr,Ti)O$_3$		Effective lattice		Th. Exp.
Material	Structure	a [Å]	c [Å]	a^* (RT)	a^*(485°C)	[ppm/K]
SrTiO$_3$	Cubic, Perovskite	3.905	-	3.937	3.967	11.4
MgO	Cubic	4.213	-	3.949	3.972	14.5
Pb(Zr$_{0.4}$,Ti$_{0.6}$)O$_3$	Tetrag. Pervskite	4.0	4.13	-	-	7.5

Table 2.1: Material properties of SrTiO$_3$, MgO and Pb(Zr$_{0.4}$,Ti$_{0.6}$)O$_3$.[218,219] The effective lattice parameter is indicated with respect to PbTiO$_3$ depositions on the respective substrate.

700°C. This means that the film 'sees' an effective lattice parameter of the substrate which is closer to the parameter of the film (see Table 2.1, p.32). The strain due to lattice mismatch can be reduced by 98% in that way.[198] The effective lattice parameter a^* is calculated as follows:

$$a^* = a(T)\,(1 - \rho(T)\,b\,)$$

$a(T)$ is the real constant of the substrate, $\rho(T)$ the linear density of primary misfit dislocations at the interface, and b is the component of the Burgers vector parallel to the film/substrate interface. In the case of PbTiO$_3$ on MgO, the effective MgO lattice at the phase transformation temperature (485°C) was calculated to 3.949Å (see Table 2.1, p.32). The a-axis of PbTiO$_3$ at that temperature is $a = 3.953$Å and the c-axis $c = 4.019$Å.[206] This means that there still exit tensile stress in the PbTiO$_3$. Just below the transition temperature this stress is released by accommodating the larger c-axis, which fits better the lattice parameter of MgO, into the growth plane, and almost complete a-axis orientation occurs. Cooling down to room temperature leads to the development of c-axis orientation. The higher thermal expansion coefficient of MgO (see Table 2.1, p.32) induces in-plane compressive stresses in the PbTiO$_3$ film, which can be released by accommodation of the smaller a-axis in the growth plane. Moreover, the c-axis of PbTiO$_3$ increases with decreasing temperature which further favors out of plane c-axis orientation.[220] The polydomain formation in epitaxial thin films has found to be independent of the depolarizing field.[198] The final fraction of a-domains was decreased to 27% at room temperature. The remaining a-domains were tilted with respect to the c-axis by 2.7° (theoretically 2.9° in PbTiO$_3$). This is a result of the accommodation of a-domains in a c-axis matrix. The theoretical tilt is illustrated in Figure 2.10(b), p.33. If the same experiment is conducted with an intermediate Pt layer, almost exclusively c-domains exist at room temperature.[194] In contrast to PbTiO$_3$ films grown directly on the MgO substrate, the epitaxial films grown on Pt/MgO experience compressive misfit strain at the growth temperature due to the smaller lattice constant of Pt. At the Curie temperature, the PbTiO$_3$ film transforms almost completely to c-axis oriented structures due to the large compressive stress induced by the substrate. Upon further decrease of the temperature, some c-domains transformed into a-domains to relax the tensile stress which developed due to the increased tetragonality of the film.

Similar results were observed on SrTiO$_3$. Considering only the room temperature situation, one would expect only c-axis oriented Pb(Zr,Ti)O$_3$ growth, since the lattice parameter of the substrate is smaller then the a-axis of PZT. The observed domain structure is characterized by $c/a/c/a$... domain patterns with a c-domain fraction of 85%.[198] The elastic energy of heterostructures is reduced by the formation of polydomain structures. Alpay et al.[198,221] studied the polydomain formation in PbTiO$_3$ on SrTiO$_3$ and MgO (001) as a function of different substrates and buffer layers. A domain stability map of PbTiO$_3$ on SrTiO$_3$ and MgO (001) was established (see Figure 2.11, p.34) for different misfit strains. The fraction of c-domains increases with decreasing PbTiO$_3$ film thickness on SrTiO$_3$ and on MgO, reaching 100% on SrTiO$_3$ for

2.4. PB(ZR,TI)O$_3$ ON SINGLE CRYSTALLINE SUBSTRATES

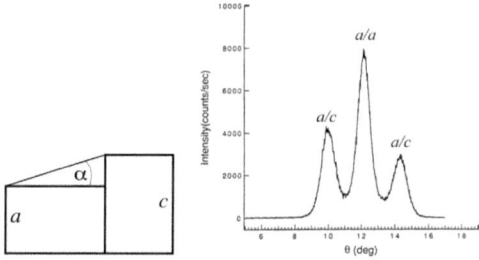

Figure 2.10: (a) Illustration of the tilt angle α of the planes from a-domains with respect to c-domains (α). For a Pb(Zr$_{0.4}$,Ti$_{0.6}$)O$_3$ the angle is calculated as 1.8°, (b) XRD rocking curve around the (100) planes of PbTiO$_3$ on (001) MgO.[191]

films thinner than 50nm. Figure 2.12(a) shows the domain structure of PbZr$_{0.2}$Ti$_{0.8}$O$_3$ obtained on LaAlO$_3$. The same domain patterns were observed earlier in single crystalline BaTiO$_3$.[125, 222] PbTiO$_3$ on (001) SrTiO$_3$ also showed mainly c-axis orientation with small a stripes running from the bottom to the top at 45°. In order not to build surface charges on the domain wall, the polarization vectors of adjacent 90°domains point in directions shown in Figure 2.12(b), p.35.

2.4.2 Influence of Film Thickness on Domains

For PbTiO$_3$ depositions on MgO/Pt, the c-domain abundance is decreased for thicker films.[194] The same was observed on SrTiO$_3$.[224] Hsu et al.[225] found an absence of a-domains on SrTiO$_3$ if the PbTiO$_3$ film was less than 50nm. Up to 25% a-domains were observed in films thicker than 350nm. In lattice-matched epitaxy the film may relax in the surface normal direction. The elastic constraint is achieved by matching the different lattice parameters[212] and relaxation is only possible near discontinuities such as domain boundaries. The concept of critical film thickness for domain formation is developed. The relative coherency e_r is defined as the ratio of misfit to tetragonality strain: $e_r = \frac{b-a}{c-a}$, with b (substrate lattice), a, c (film lattices). Below a critical film thickness, perfect c-orientation is obtained. Thicker films grow c-axis oriented if e_r is small enough, and a-axis oriented for large e_r. In between, mixed orientations will be obtained.

Lee et al.[201] used PLD of PbTiO$_3$ on MgO (001) at 700°C to test experimentally the thickness-dependent relaxation of elastic misfit strains on domain formation. The a-domain structure was observed by synchrotron measurements as a function of thickness: As film thickness increases, the elastic coherency strain energy increased linearly with the thickness, and eventually exceeded the energy required for misfit dislocation generation. As the thickness increased, the dislocation density rapidly increased and a marked thickness dependence of the misfit strain was observed at about 100nm. Above 100nm the effective misfit strain decreased only slowly and approached a constant value even for an infinite film thickness.[201] A steep increase was observed up to 100nm, after that, the abundance reaches slowly 75%.

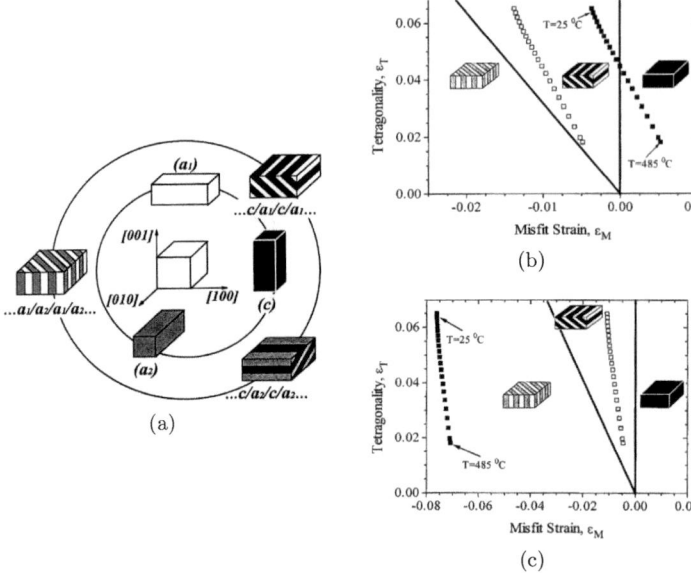

Figure 2.11: Domain stability maps[216, 221] of epitaxial PbTiO$_3$ on SrTiO$_3$ and MgO (100) as a function of misfit strain. (a) the different domain configuration possibilities, (b) PbTiO$_3$ on SrTiO$_3$, (c) PbTiO$_3$ on MgO. Open squares show data for a film completely relaxed by misfit dislocations at the deposition temperature, solid squares for an unrelaxed film.

2.4.3 Domain Spacing

Depending on the substrate, and on the film thickness, the density and the mean width of a-domains varies.[226] On SrTiO$_3$ the a-domain width of 500nm thick PbTiO$_3$ was 30nm, spacing 300nm, whereas on MgO (100) the width was 30nm, spacing 160nm, and on SrRuO$_3$ buffered SrTiO$_3$ the width was 42nm, spacing 250nm.[226]
It is theoretically found that the domain wall energy γ increases with decreasing domain width t_d as $\gamma \propto \frac{P^2}{t_d}$ (P is the polarization). The domain width at room temperature decreases linearly with the particle size. For a given particle size, the domain width increases with decreasing space charge layer thickness.[101] The free energy of the 180° domain width as a function of domain width (for a given film thickness) was found to have a minimum.[104] This preferred domain width w decreases with decreasing film thickness d as $w \propto \sqrt{d}$. This is also found in ferromagnetics.[227] Experimentally, Mitsui et al.[135] found this behavior in rochelle salt. They found a steep increase to infinity for the domain density with shrinking film thickness. The domain width was decreased with film thickness following the above stated relation.
The size dependent ferroelectric domain structure in free-standing PbTiO$_3$ thin films composed of 60 to 100nm large grains was studied[27, 228] by means of TEM. PbTiO$_3$ was deposited onto NaCl and NaCl was dissolved afterwards. The fine grain size was obtained by heating locally

2.4. PB(ZR,TI)O$_3$ ON SINGLE CRYSTALLINE SUBSTRATES

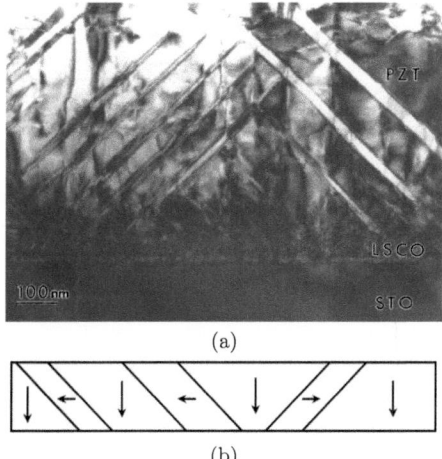

(a)

(b)

Figure 2.12: (a) Domain structure in PbZr$_{0.2}$Ti$_{0.8}$O$_3$ film on LaAlO$_3$ with LSCO buffer layer.[223] The a-domain width has an average of $30nm$ spaced by $100nm$. (b) schematic of the polarization directions. To avoid charged domain walls, the polarization vector's projection of neighboring domains on the domain wall's normal have to point to the same direction.

the film in the electron beam. A transition from predominantly multidomain to single domain occurs at a grain size of about $150nm$. The grain size, rather than film thickness is the controlling factor determining the domain structure and the electrical properties. For $1\mu m$ large grains, the domain width was $70-200nm$. For $150nm$ large grains, the domain width was $10-50nm$, and domain patterns were rarely seen in grains smaller than $60nm$. At even smaller sizes, single domain grains became stable and dominating. Thus, the observed thickness dependence is due to the grain size, as the thickness of the film is related to the grain size. The polarization change due to domain wall motion at low field is insignificant and consequently the associated permittivity is low due to the lack of domain walls in thin films. The sharp increase in coercive field at smaller sizes can be explained by the stable and dominating single domain grains which make nucleation of domains difficult. Thin films down to $80nm$ were also investigated.[228] Single domained grains show an unexpected resistance against the formation of new domains.

On the one hand the domain width decreases with decreasing feature size, and on the other hand, the domain wall energy increases with decreasing domain width. It can be expected, that for small sizes there can't exist a-domains any more. Restrictions in the domain configuration can be expected upon downscaling (see Figure 2.13, p.36).

2.4.4 The Effect of Seed Layers

The microstructure of a thin film, such as grain boundaries, grain size and crystallographic orientation influence greatly it's physical properties. The microstructure depends above all on the nucleation and growth of the film. The substrate's surface influences the nucleation density,

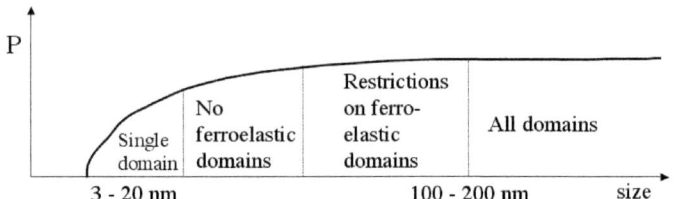

Figure 2.13: Illustration of the restriction on the domain configuration upon downscaling. At large sizes, all domains are possible. At smaller sizes d, the domain width w shrinks with $\propto \sqrt{d}$, at the same time, the increased domain wall energy makes their presence less probable. Before the complete vanishment of the polarization due to size effect, a monodomain state is expected.

growth mode, and orientation of the film. Seed layers can help to promote a desired parameter, such as high nucleation density and orientation. A surface modified by a surfactant may promote layer-by-layer growth.[229] As an example, a monolayer of Pb on Cu acts as surfactant layer. During surfactant-assisted epitaxial growth[230] the Cu atoms deposited onto the surface get buried below Pb[231] and join the underlying Cu. For magnetic recording media devices, $2nm$ Ag seed layers increased the coercive field of CoCrPt/Ti thin films[232] and smaller domains were obtained. An $SrTiO_3$ seed layer on MgO (100) expanded the range of deposition temperature of YBaCuO by $60°C$. The influence of Ta seed layers on the ferroelectric properties of $SrBi_2Ta_2O_9$ were also investigated.[233] Higher nucleation density and higher remanent polarization were found for depositions on a Ta seed layer. Patterning of a nucleation density enhancing seed layer can be used for site-selective growth. Self assembled monolayers (SAMs) can also play the role of a seed layer for site-selective depositions. SAMs have been used to tailor the chemical characteristics of substrate surfaces by producing highly specific interfaces spread over relatively large areas, which also can be modified chemically or physically to generate surfaces having different reactivity and properties. A commonly used method is UV-irradiation through a photomask.[234] The modified SAMs can then be employed as templates to fabricate micro-patterns of functional materials and small devices of the required size and shape with outstanding precision through a one-step process at room temperature. However, the deposition takes place in a liquid phase, like in the sol-gel method, and the formation of crystalline material requires often a post-firing step. Site-selective deposition of tin oxide (SnO_2) has been prepared on a patterned self assembled mono-layer[235] using molecular recognition between the precursors and the surface of an Si substrate. UV irradiated regions became hydrophilic through functional group change. The tin precursors only deposited onto the hydrophilic regions. Site-selective deposition was also achieved to fabricate $SrTiO_3$ micro-patterns combining liquid phase deposition (LDP) and SAM. UV-irradiation of the SAM led to hydrophilic silanol (Si-OH) patterns where LDP took place. $SrTiO_3$ was crystallized by post-annealing at $500°C$.

Using again SAMs and UV irradiation amorphous TiO_2 seed layers were deposited from the liquid phase on the irradiated SAMs. Subsequent TiO_2 deposition from a liquid phase led to the formation of crystalline anatase TiO_2 on the amorphous TiO_2. The amorphous TiO_2 seed layer accelerated the growth of anatase TiO_2.[236]

AlN was used as a nucleation seed layer on SiC for the selective growth of GaN.[56] Columnar growth was observed on AlN seeds and the lateral overgrowth led to smooth facets on the sides of

2.4. PB(ZR,TI)O₃ ON SINGLE CRYSTALLINE SUBSTRATES

Figure 2.14: Selective growth of GaN on AlN seed stripes. In between the stripes, the SiC substrate can be seen, where no nucleation took place.[56]

the GaN stripes. No deposition on SiC was observed in the $3\mu m$ gap between two seed regions[56] (see Figure 2.14, p.37). In this work, selective nucleation of Pb(Zr,Ti)O₃ on TiO₂ seeds will be shown.

2.4.5 Seed Layers for Pb(Zr,Ti)O₃

For Pb(Zr,Ti)O₃, the control over the polarization vector's orientation of thin oxide ferroelectric films and the growth mode is of great importance to obtain the desired properties.[237–239] The growth mode is dependent on the relative surface free energies of the substrate and the deposited film, and/or on the oxygen affinity between the arriving atoms and the surface. It was found, that nucleation of Pb(Zr,Ti)O₃ on Pt was difficult. Direct deposition of Pb(Zr,Ti)O₃ on platinized Si substrates exhibited mixed perovskite/pyrochlore phases.[240] This might be due to the low oxygen affinity of Pt. TiO₂ proved to be a good adhesion layer for Pb(Zr,Ti)O₃ because it wets well all kind of surfaces due to its low surface energy, and continuous thin layers ($2nm$) can be deposited. Moreover, TiO₂ was shown to be a promoter or dense nucleation and (111) growth of PZT. This must be attributed to the high oxygen affinity. Depositions of fcc metals on TiO₂ led to island growth,[241] as expected for depositions of high surface energy material like Pt[242] on a low surface energy surface. Pb(Zr,Ti)O₃ depositions have been studied extensively using a large number of deposition methods on various substrates. Still the most important combination is a substrate with inert Pt electrodes. The main problems for Pb(Zr,Ti)O₃ depositions on a Pt electrode is the formation of pyrochlore phases and interdiffusion of Pb species through the electrode into the substrate. Barrier layers such as SiO₂, Ti or TiN are required.[239] Stoichiometric Pb(Zr,Ti)O₃ depositions on oriented Pt (111) often lead to mixed perovskite of the orientations (110), (001), (100) and (111), and to pyrochlore phases. The pyrochlore phase is transformed into perovskite by an annealing step after deposition. Pure platinum is not the ideal substrate to nucleate the perovskite phase.[243] A low nucleation density, and therefore high leakage currents are observed. It is generally observed, that Zr rich films typically show a microstructure with large perovskite grains, often called rosettes.[55] Ti rich layers have much smaller grain sizes, indicating higher nucleation rates.
Pb(Zr,Ti)O₃ has been grown epitaxially on single crystaline Pt (111) and (100) (see Figure

3.27, p.70), but island growth with poor surface coverage occurs for stoichiometric depositions. Rhombohedral Zr rich Pb(Zr,Ti)O$_3$ was epitaxially grown on epitaxial Pt (111) on single crystalline sapphire (0001) substrates using a target with excess PbO.[244] On (100) Pt an underlying unknown layer forms, and X-ray analysis show small amounts of other phases. Epitaxial effects, *i.e.* the small lattice mismatch between Pt an tetragonal Pb(Zr,Ti)O$_3$ compositions (on which we are focussing) is not the only driving force for orientation of Pb(Zr,Ti)O$_3$ thin films. In sol-gel depositions, the application of a $2nm$ thick Ti film on Pt electrode greatly enhanced the nucleation density of (111) oriented PZT[243] and resulted in fine grained Pb(Zr,Ti)O$_3$ with low leakage currents. (100) oriented PbZr$_{0.63}$Ti$_{0.37}$O$_3$ were obtained on Si/SiO$_2$/Ti/Pt substrates, but after applying a TiO$_2$ seed layer before Pb(Zr,Ti)O$_3$ deposition, (111) growth was obtained.[238] On the other hand, sol-gel prepared $100nm$ thick PbTiO$_3$ template layers were found to improve the growth of the perovskite phase.[245]

Pb(Zr$_{0.53}$Ti$_{0.47}$)O$_3$ films on Si/SiO$_2$/Pt using PbTiO$_3$ seed layer were prepared by sol-gel.[246] The PbTiO$_3$ seed layer was $30nm$ thick and induced (001) orientation of the Pb(Zr,Ti)O$_3$ film. Lead rich PbTiO$_3$ preferentially grows in the (001) direction, probably due to the low surface energy of PbO terminated (001) surfaces.

In sputtering systems with adjustable Pb, Ti, Zr flows[210] it was shown, that a high Pb flow leads to denser nucleation of (100) PbTiO$_3$ growth on oriented Pt (111).[247] The orientation of Pb(Zr,Ti)O$_3$ on oriented (111) Pt electrodes can be controlled using seed layers of the type (TiO$_2$)$_x$(PbO)$_{(1-x)}$. Subsequent Pb(Zr,Ti)O$_3$ deposition on Pb rich seed layers led to (100) Pb(Zr,Ti)O$_3$, on Ti rich seed layers, very dense (111) oriented Pb(Zr,Ti)O$_3$ was obtained[247] even for very thin TiO$_2$. For an equivalent TiO$_2$ thickness of the (TiO$_2$)$_x$(PbO)$_{(1-x)}$ seed layer of $2nm$, the Pb(Zr,Ti)O$_3$ changed the orientation from (111) to mixed (100), (111) and (110) when the PbO flux is more than 2.2 times higher than the TiO$_2$ flux (see Figure 2.15, p.39). Further increasing the PbO flux to above 3.3 times with respect to the TiO$_2$ flux completely oriented the sputtered Pb(Zr,Ti)O$_3$ to (001) (c-axis).[247] TiO$_2$ seed layers inevitably led to a (111) orientation, independent of the Pb(Zr,Ti)O$_3$ deposition conditions.[247] Without applying a seed layer, c-axis oriented Pb(Zr,Ti)O$_3$ was obtained on (111) oriented Pt for lead oxide fluxes exceeding twice the flux of TiO$_2$. The pyrochlore phase was obtained for lower lead fluxes. With TiO$_2$ seed layer (111) growth occurred independent of the lead flux.[248] The crystalline structure of TiO$_2$ is emphasized, as cold deposited TiO$_2$ seeding layers did not have an orientation impact on PZT. $100nm$ TiO$_2$ on oriented Pt (111) exhibited random rutile-anatase orientation. The surface energy of PbO is supposed to be the smallest because of the small bonding energy between PbO. The reason for (001) (c-axis) growth is explained in terms of surface and interface energies. For high PbO fluxes, a low surface energy (PbO, litharge, layer structure) favors TiO$_2$/PbO sequences, which correspond to the Pb(Zr,Ti)O$_3$ (100) growth. In the case of (111) depositions epitaxial considerations are thought to be the driving force for the orientation.

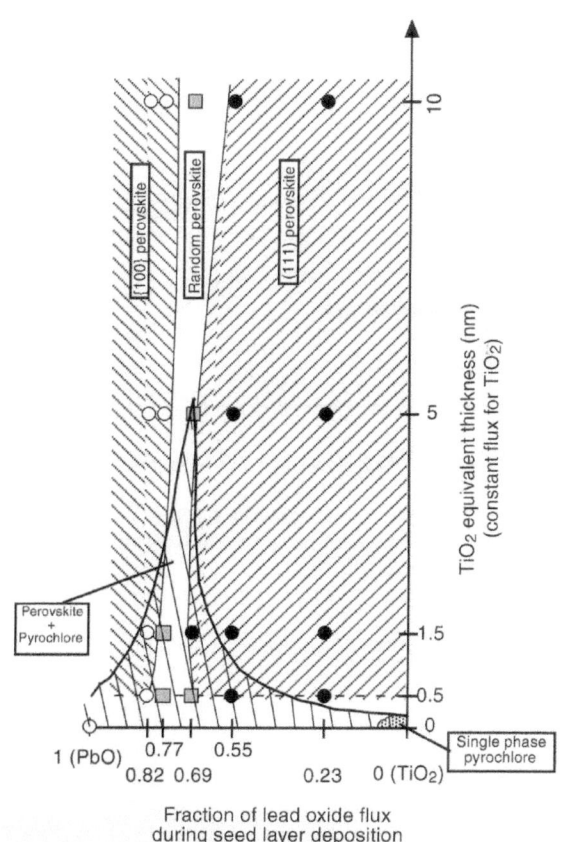

Figure 2.15: Diagram[247] summarizing Pb(Zr,Ti)O$_3$ texture and presence of pyrochlore as a function of the seed layer composition (TiO$_2$)$_x$(PbO)$_{(1-x)}$ and thickness on (111) oriented Pt on Si/SiO$_2$.

Chapter 3

Growth of Epitaxial Thin Films

In this chapter, the deposition of the thin films used in the subtractive and additive routes are discussed. The analysis are done on the non-patterned continuous films. The crystallographic orientation of the films is analyzed by XRD pole figure measurements.
First, the deposition techniques are explained and a short introduction to X-ray diffraction is given highlighting the use of pole-figures to determine the crystallographic in-plane relation between layers. The following thin film systems are presented:

For the subtractive route the deposition of Pb(Zr,Ti)O_3 on Nb-doped SrTiO_3 (100) is discussed and compared with depositions of Pb(Zr,Ti)O_3 on SrTiO_3 (111). For a continuous (001) Pb(Zr,Ti)O_3 film on SrTiO_3 (100) polarization loops and d_{33} measurements are shown and the Landau parameters calculated to know about critical thicknesses.

For the additive route two thin film systems are presented:

- TiO_2/ Pt/SrTiO_3 (111)
- TiO_2/Pt/MgO (100)

The direct deposition of Pt on MgO and SrTiO_3 (100) and Pt depositions using an interlayer of Ir are discussed. For the additive route it is of particular interest to know the orientation of the TiO_2 seeds with respect to Pt. This is investigated using $200nm$ thick TiO_2 on Pt (100) and (111). For the deposition of Pb(Zr,Ti)O_3, several investigations were made: As we want to use the different nucleation behavior of Pb(Zr,Ti)O_3 on bare Pt and TiO_2 covered surfaces in the additive route, the results of direct Pb(Zr$_{0.4}$,Ti$_{0.6}$)O_3 depositions on Pt are presented, and depositions with a PbTiO_3 layer prior to Pb(Zr,Ti)O_3 deposition on Pt (100) and (111) are shown.

3.1 Deposition of Thin Films

For the deposition of thin films we use the following methods:

- Sputtering (Ir, Pt, Pb(Zr,Ti)O$_3$, TiO$_2$, PbTiO$_3$)
- Evaporation (Cr hard mask)
- Spin Coating (Electron beam resists: PMMA, Calixarene)

3.1.1 Sputtering

The sputtering method is used to deposit epitaxial films of:

- Pb(Zr,Ti)O$_3$
- PbTiO$_3$
- TiO$_2$
- Pt
- Ir

on various single crystalline substrates or onto epitaxial layers. The deposition conditions can be found in Table 3.1, p.43. A Nordiko 2000 sputtering system was used for thin film deposition. A primary rotary and a secondary high vacuum turbo pump combined with a liquid N$_2$ Meissner trap allow to pump to a base pressure of $3 \times 10^{-6} mbar$ measured above the Meissner trap. The chamber features a rotatable substrate holder table which is mounted upside-down. The substrate holder could be heated to up to $800°C$. This was the nominal temperature measured inside the heated chuck. On the surface of the sample, the temperature was lower. A nominal temperature of $700°C$ corresponded to a surface temperature of $600°C$. The substrate holder was rotated at a variable distance from the target table, witch provided positions for three distinct water cooled targets with magnetrons. The magnetic field of the magnetron was oriented parallel to the cathode (target) surface. Due to the increased confinement of the secondary electrons in this external magnetic field, the plasma density is much higher and can be sustained at significantly lower chamber pressures. The used pressures were in the range from 4 to $20 mTorr$. Two targets could be operated in DC mode, one at radio-frequency (RF, $13.56 MHz$). Two gas inlets were used: O$_2$ and Ar. The pressure was sensed by a Baratron and controlled by means of a plate valve.

For the Pb(Zr,Ti)O$_3$ process, the heated substrate was rotated over single metal targets of Pb, Ti and Zr in an oxygen atmosphere at $16 mTorr$. The lead target was operated with an RF generator, the others worked in DC mode. The rotation speed of the substrate was $6 rpm$. This process corresponded to a pulsed deposition (see Figure 3.1, p.44), which was important in the additive route, where diffusion issues played a great role. For a rotation at a radius $r = 0.3m$ and a supposed interaction with the plasma above each target (diameter $10 cm$) of $20 cm$, each oxide was deposited during a pulse time of roughly $1s$ onto the substrate (adsorption/desorption of oxide species with arriving material). Between Zr–Pb and Pb–Ti, the substrate ran through a zone without deposition during $1.5s$ each, and between Ti–Zr during $4s$ (adsorption/desorption of oxide species without arriving material). This corresponded to a deposition of roughly $1 ML$ per turn.

3.1. DEPOSITION OF THIN FILMS

#	Material	O$_2$ [sccm]	Ar [sccm]	p [mT]	T [°C]	rot [rpm]	Pt/Ir DC1 [W]	Pb RF1 [W]	Ti DC2 [W]	Zr DC4 [W]	Rate [nm/min]
[1]	TiO$_2$	20	-	10	650	static	-	-	100	-	1.3
[2]	PbTiO$_3$	20	-	16	700	6	-	200	490	-	3
[3]	Pb(Zr$_{0.4}$,Ti$_{0.6}$)O$_3$	20	-	16	700	6	-	150	310	128	3
[4]	Pb(Zr$_{0.1}$,Ti$_{0.9}$)O$_3$	20	-	16	700	6	-	150	445	32	3
[5]	Ir	-	30	5	700	static	20	-	-	-	8
[6]	Pt	-	30	4	700	static	-	30	-	-	2.3
[7]	Pt	-	30	10	400	static	50	-	-	-	16.6

Table 3.1: Sputtering deposition conditions.

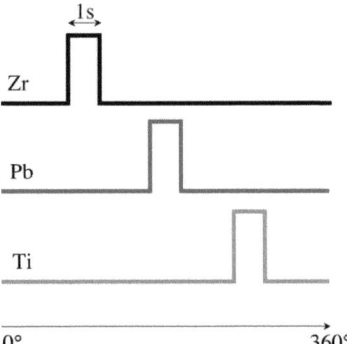

Figure 3.1: Schematic of the sputtering process used for Pb(Zr,Ti)O$_3$ deposition. The substrate is rotated above the targets and is submitted to deposition pulses of the oxide species from the single metallic targets.

3.2 X-Ray Diffraction

One characterization of the deposited crystalline phase of thin films is X-ray diffraction.[249] We use $\theta - 2\theta$ scans, rocking curves and pole figure measurements as described below.

3.2.1 $\theta - 2\theta$ Scans

In this measurement the detector and the sample surface are scanned relatively to the fixed source in a way that the surface normal always cuts the angle between the source-sample and sample-detector in halve (*i.e.* if the sample rotates by θ, the detector has to do a 2θ step).
According to Bragg's condition for constructive interference, there is an X-ray peak intensity at the condition $2d \sin\theta = n\lambda$ (d: spacing of an atomic plane, θ: angle between source and the sample surface, $\lambda = 1.5406$Å for Cu$_\alpha$ X-ray emission). In this way, the reciprocal lattice vector $\vec{r^*}$ (see below) always stays perpendicular to the substrate plane. This means that only the crystal planes parallel to the substrate can be observed.
This measurement does not give information about the in-plane orientation of the film. It only shows its orientation with respect to the substrate normal.

3.2.2 Rocking Curves

In a rocking curve measurement, the position of a certain existing constructive diffraction is fixed and the substrate is rotated ("rocked") around this position. The width of the so acquired rocking curve is a measure of the tilting spread of the corresponding planes. This method was mainly used to characterize the perfection of single crystalline materials.[249]

3.3. SRTIO₃ SUBSTRATE PREPARATION

Cubic Crystals

	[110]	[101]	[111]
[001]	90°	45°	54°
[110]	×	60°	35°
[101]	60°	×	35°

(a)

Rutile TiO₂

	[100]	[110]	[101]	[111]
[001]	90°	90°	33°	42°
[100]	×	45°	57°	61°
[110]	45°	×	67°	47°
[101]	57°	67°	×	28°

(b)

Table 3.2: Angles between different directions in (a) cubic crystals and (b) rutile TiO₂ crystal.

3.2.3 Pole-Figure Measurement

In the case of pole figure measurement the sample can be rotated around 3 axis as indicated in Figure 3.2, p.46. In our case, the source is fixed. For a standard $\theta - 2\theta$ scan the relative rotations of the detector and sample (ω) take place. For a pole figure measurement of a certain atomic plane, the Bragg reflection is fixed by the 2θ angle between the detector and the source, and the sample is rotated around χ and ϕ. Whenever the reciprocal vector of the investigated atomic plane points in the center of the angle source-sample-detector, a constructive interference takes place and a peak is detected. This measurement shows the in-plane relation of a film with the underlying substrate. The data is plotted in cylindrical coordinates with [radius, angle]=$[\chi, \phi]$. In this work we deal with quasi cubic perovskite structure and rutile TiO₂. To facilitate the readability of the polefigures shown, Table 3.2, p.45 shows the angles (χ) between two atomic plane's reciprocal vectors. The calculation within any crystalline system can be done as follows[250] (* designates the reciprocal space):
The reciprocal vector $\vec{r^*}$ has the same direction as the normal $\vec{N}$ to the atomic plane given by the Miller indices khl. $\vec{r^*}$ is given by:

$$\vec{r^*} = h\vec{a^*} + k\vec{b^*} + l\vec{c^*} = \frac{\vec{N}}{d|\vec{N}|} \quad d: \text{spacing of planes}$$

with

$$\vec{a^*} = \frac{\vec{b} \times \vec{c}}{(abc)}, \quad \vec{b^*} = \frac{\vec{c} \times \vec{a}}{(abc)}, \quad \vec{c^*} = \frac{\vec{a} \times \vec{b}}{(abc)} \quad \text{and} \quad (abc) = \vec{a} \cdot (\vec{b} \times \vec{c})$$

where

$\vec{a}, \vec{b}, \vec{c}$ are direct space unit cell vectors, known from crystallographic data

From these formulas, any angle between crystal planes can be calculated in the direct space. This is helpful for the pole figure measurements, where the angle between the surface normal (in our work the [100] or [111] direction of a single crystal MgO or SrTiO₃) and another direction has to be calculated. Table 3.2, p.45 gives some examples for angles between some directions in cubic crystals (a) and in the tetragonal rutile TiO₂ crystal (b).

3.3 SrTiO₃ Substrate Preparation

To achieve epitaxy, the atomic arrangement of the topmost few layers are of great importance for the orientation of the subsequent material. In the case of SrTiO₃ (100) the topmost layer

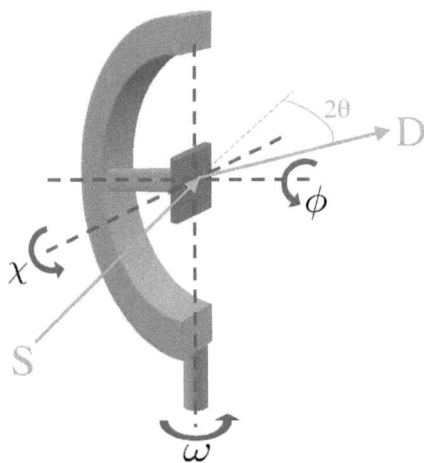

Figure 3.2: Setup for polefigure measurements showing X-ray source (S, fixed), detector (D), the three axis of possible sample rotation χ, ω, ϕ, and the Bragg reflection angle 2θ.

can be SrO or TiO_2. Figure 3.3(a), p.47 shows the stacking of atoms in $SrTiO_3$ with a shaded (100) layer of TiO_2. For $SrTiO_3$ (111) this layer can be SrO_2 or Ti. Figure 3.3(b), p.47 shows a (111) plane of Ti atoms. The surfaces of $SrTiO_3$ (111)[251] and (100)[252–254] have been investigated extensively in literature. $SrTiO_3$ surfaces are of interest not only for epitaxial atomic layer depositions of perovskite oxides[255–257] but also for the photocatalysis of the water splitting reaction.[253,254] It was generally found, that polishing an $SrTiO_3$ (100) surface led to mixed SrO and TiO_2 terminated regions[258] with roughly 25% SrO. For epitaxial depositions, single terminated surfaces are required.[259] On $SrTiO_3$ (100) surfaces, it was shown that single terminated TiO_2 surfaces could be obtained by selectively etching the basic oxide SrO or by a Bi-deposition / desorption treatment followed by a heat treatment.[260] UHV high temperature annealing led to roughening of the surface,[256] Sr segregation above $700°C$[261,262] and a decrease of the TiO_2 fraction.[263] For the selective etching, the topmost SrO layer can be hydroxylated in water and then dissolved in a buffered HF (BHF) solution.[254,263–265] Hydroxylation in water was also observed on (111) surfaces.[251] We used the selective etching method in BHF for our substrates to obtain single terminated TiO_2 surfaces. The $SrTiO_3$ (100) and (111) substrates were treated for three cycles comprising a $1 min$ hydroxylation in deionized H_2O followed by a $30 s$ etching in NH_4F:HF=87.5:12.5, pH=5.5 solution.[264] Finally the samples were rinsed in H_2O.

In our work, XRD measurements revealed the importance of the surface treatment in the case of $Pb(Zr,Ti)O_3$ depositions on $SrTiO_3$ (100) and for depositions of Pt and TiO_2 on $SrTiO_3$ (111). Figure 3.4, p.48 shows the influence of the BHF-H_2O treatment on depositions of $200 nm$ $Pb(Zr,Ti)O_3$ on $SrTiO_3$ (100) revealed by XRD measurements of the a and c-axis. The measured values are plotted on the data found by Foster et al.[209] (see Figure 2.9, p.31). Extremely high lattice distortions of $Pb(Zr,Ti)O_3$ were found for Ti rich depositions (black spots) on un-

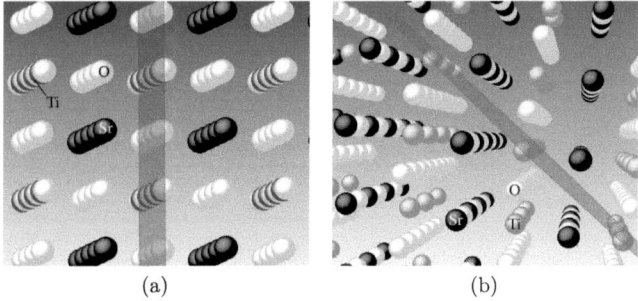

Figure 3.3: Stacking in SrTiO$_3$ along (100) and (111) directions. The corresponding plane is shaded. In the (100) direction (a) the layers are composed of SrO or TiO$_2$, and in the (111) direction (b) the layers are composed of SrO$_3$ and Ti.

treated SrTiO$_3$ (100) surfaces leading to an increase of the unit cell volume of up to 11% for a composition of 90% Ti. Treating the SrTiO$_3$ surface in BHF before Pb(Zr,Ti)O$_3$ deposition considerably reduced the lattice distortion. The in-plane a-axis was now in agreement with the values found by Foster et al.[209] (see circles in Figure 3.4, p.48). The distortion of the c-axis was also reduced but remained at a higher value of 4.2Å. This is attributed to our sputtering process which produces lead rich compositions containing 4-valent B-site lead (PbPbO$_3$) which leads to an increase of the unit cell volume.[266]

For depositions of TiO$_2$ on Pt/SrTiO$_3$ (111), the surface treatment of the SrTiO$_3$ (111) surface led to better in-plane orientation. Figure 3.5(a), p.48 shows the pole-figure of the [110] direction of a 200nm thick TiO$_2$ layer on Pt/SrTiO$_3$ (111) without BHF treatment of the SrTiO$_3$ surface. The measured peaks are located at 45°, indicating TiO$_2$ (100) growth. (a) shows the result without BHF treatment. Compared to substrates with BHF treatment (b), the in-plane orientation in (a) is smeared out over a wide ϕ-angle (about 30°). This was induced by the Pt layer which showed similar in-plane orientation.

The initial growth stage of MOCVD deposited Pb(Zr,Ti)O$_3$ on TiO$_2$ and SrO terminated SrTiO$_3$ (100) surfaces was studied by Fujisawa et al.[58, 267] Friction force AFM measurements on SrTiO$_3$ (100) surfaces revealed high friction SrO and low friction TiO$_2$ terminated terraces. Pb(Zr,Ti)O$_3$ showed the Volmer-Weber growth mode on the SrO terminated surface and Stranski-Krastanov mode on the TiO$_2$ terminated surface. PbO depositions showed Volmer-Weber island growth on the SrO terminated surface and layer by layer Strankski-Krastanov growth with a critical thickness on less than 10 unit cells on the TiO$_2$ terminated surface. The deposition of TiO$_2$ was layer-by-layer on both terminations. Therefore, Pb(Zr,Ti)O$_3$ growth on the SrO surface starts with a PbO layer, and on the TiO$_2$ terminated surface by PbO or TiO$_2$.

3.4 Pb(Zr,Ti)O$_3$ on SrTiO$_3$ (100)

In this section, direct depositions of tetragonal Pb(Zr,Ti)O$_3$ compositions on SrTiO$_3$ (100) will be discussed. These films are used in the subtractive route. For the interpretation of results, it is important to know about the domain structure of the continuous film. This was done by XRD diffraction and TEM analysis. Moreover, the piezoelectric properties of the continuous film have

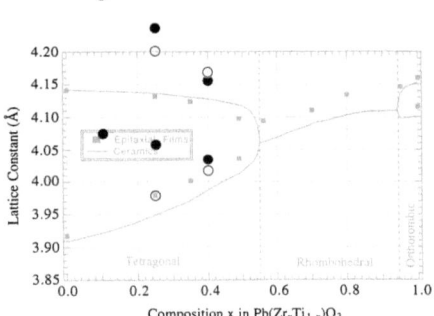

Figure 3.4: Effect of substrate preparation on the a- and c axis spacing for sputtered Pb(Zr,Ti)O$_3$ films of different compositions on SrTiO$_3$ (100). • unprepared, ○ BHF treated. The values were measured by means of X-ray diffraction and are shown superimposed on the values found in the literature.[209]

been measured. Ferroelectric small-signal d_{33} loops have been measured by an interferometer, and polarization loops have been acquired as well. From these data, and using the LDG theory (see p.14), it was possible to calculate the critical film thickness, below which the polar phase is unstable at all temperatures.

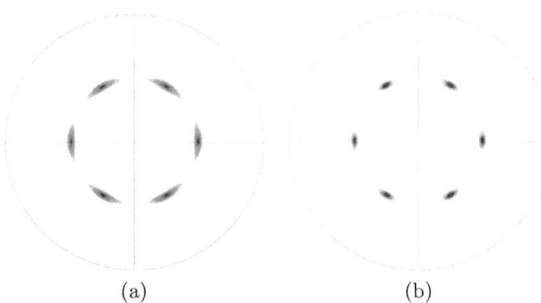

Figure 3.5: Pole-figure measurement on $200 nm$ TiO$_2$ on $50 nm$ Pt on SrTiO$_3$ (111). The measured direction is [110] at $2\theta = 27°.446$. The peak position at $\chi = 45°$ indicates TiO$_2$ (100) growth (see also Table 3.2, p.45). (a) without cleaning the SrTiO$_3$ substrate in BHF, (b) including treatment of the SrTiO$_3$ (111) substrate in BHF (NH$_4$F:HF=87.5:12.5, pH=5.5) and water.[264]

3.4. PB(ZR,TI)O$_3$ ON SRTIO$_3$ (100)

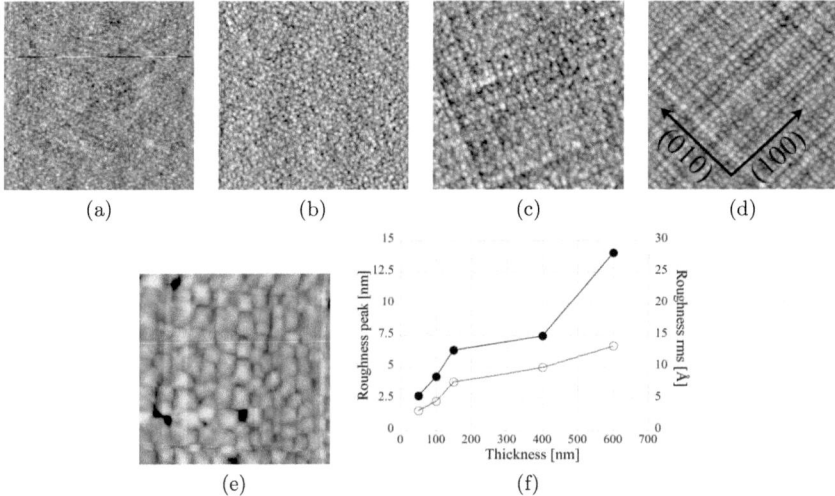

Figure 3.6: AFM-topographies of Pb(Zr$_{0.4}$,Ti$_{0.6}$)O$_3$ films on SrTiO$_3$ (100) single crystalline substrate. (a) 50nm, (b) 100nm (c) 150nm (d) 400nm (e) 600nm. The images show 5×5μm scans. (f) Roughness of Pb(Zr$_{0.4}$,Ti$_{0.6}$)O$_3$ on SrTiO$_3$ (100) as a function of film thickness measured by means of AFM. • peak to peak roughness, ○ root mean square (rms) roughness.

3.4.1 X-Ray Characterization

For the subtractive route, Pb(Zr,Ti)O$_3$ was deposited directly on Nb-doped conductive SrTiO$_3$ (100) substrates with thicknesses ranging from 50 to 600nm. Pb(Zr$_{0.4}$,Ti$_{0.6}$)O$_3$ was deposited using the conditions listed in Table 3.1, p.43. Figure 3.6, p.49 shows the topography and roughness of the Pb(Zr,Ti)O$_3$ films as a function of thickness of the Pb(Zr,Ti)O$_3$ film measured by AFM in contact mode. Up to a thickness of 100nm random surface pattern can be observed. Above this thickness, an oriented pattern starts to appear which follows the crystallographic in-plane [100] and [010] orientation of the substrate (indicated in (d)). Dislocations are not restricted to the interface any more and can extend over the film thickness (threading dislocations).[146, 150] This generates the pattern observed above 100nm which is the critical thickness for dislocation formation in the thin film. Such pattern were also observed on BaTiO$_3$ films on SrTiO$_3$ (100) single crystals.[268, 269] Up to 400nm thick films, the root mean square roughness increases in the same way as the peak to peak roughness. For 600nm thick films, the presence of few nm deep holes led to a sharp increase in the peak to peak roughness. XRD $\theta - 2\theta$ scans showed a dominant c-axis orientation Pb(Zr,Ti)O$_3$, with some minor amount of a-axis (100) orientation (see Figure 3.7(a), p.51). Pole figure measurements of the Pb(Zr,Ti)O$_3$ and SrTiO$_3$ [011] directions confirmed the epitaxial growth (b). The larger gray spots arise from Pb(Zr,Ti)O$_3$, and in the center the black line corresponds to the SrTiO$_3$ [011]. The lattice parameters of the Pb(Zr,Ti)O$_3$ was found to be dependent on the thickness of the films (see Figure 3.8, p.51). The graphs (a–c) show the measured spacings of Pb(Zr,Ti)O$_3$ (002) (a), (011) (b) and (111) (c). The c-axis and the (001) spacing increased with decreasing film thickness while the (111) spacing showed an initial increase and then a sharp drop at a thickness of $50nm$. (d) shows the calculated in-plane (200) spacing of Pb(Zr,Ti)O$_3$, based on the tetragonal assumption. This spacing was calculated in three ways: Using the data from (002) and (011), (002) and (111), and (011) and (111). All calculated values are in good agreement with each other. Due to clamping, thinner films show smaller Pb(Zr,Ti)O$_3$ (200) values, approaching the value of SrTiO$_3$ (200) ($d_{(200),SrTiO_3}$ = 1.953Å). For thicker films, the length of the a-axis stabilized at 4.054Å. The volume of the unit cell is constant. At $400nm$ the unit cell volume is 68.533Å^3. To keep this volume constant, a c-axis value of 4.232Å would be required for an a-axis of 4.024Å (for $50nm$ thin film). The measured c-axis is 4.234Å which is in good agreement with the hypothesis of constant volume.

In Figure 3.7, p.51 an additional peak appears between Pb(Zr,Ti)O$_3$ and SrTiO$_3$ (002) which corresponds to Pb(Zr,Ti)O$_3$ (200) (a-domains). Rocking curves performed around this direction revealed the tilting of a-domains with respect to the normal direction of the surface by 1.8° (see Figure 3.9, p.52), which corresponds to the theoretical value (see Figure 2.10, p.33). The amount of a-domains in Pb(Zr$_{0.4}$,Ti$_{0.6}$)O$_3$ on SrTiO$_3$ (100) can be estimated from XRD measurements. This can be done considering the maximum intensities of the a-domain configurations (a/c and a/a) from the rocking curve and the maximum intensity of the Pb(Zr$_{0.4}$,Ti$_{0.6}$)O$_3$ (001) c-domains. In three dimensions, there exist 4 a/c reflections. It is supposed that the peaks have the same full width at half maximum, and that the diffraction yield from (100) planes is the same than from (001) planes. A clear contribution from a-domains appear in $400nm$ and $600nm$ thick Pb(Zr$_{0.4}$,Ti$_{0.6}$)O$_3$ films (see Figure 3.9, p,.52). For a $600nm$ thick Pb(Zr$_{0.4}$,Ti$_{0.6}$)O$_3$ film, the a domain abundance is estimated as 25 ± 5%, and for a $400nm$ thick film, it is 14 ± 3%. This increase is a-domain abundance with increasing Pb(Zr,Ti)O$_3$ film thickness has also been observed by others.[224, 270] Thanks of the high tetragonality of 2%, we were able to observe the domains in by TEM. Figure 3.10, p.52 shows cross-section dark-field images highlighting a-domains in a $400nm$ thick Pb(Zr$_{0.4}$,Ti$_{0.6}$)O$_3$ film on SrTiO$_3$ (100). In (a), stripes as thin as $25nm$ arising from a/c domain walls can be seen. The incident angle with the substrate is 45°. An explanation for a possible domain structure has been given before (see Figure 2.12, p.35). Figure 3.10(b), p.52 shows a/a domains featuring vertical domain walls.

3.4. PB(ZR,TI)O$_3$ ON SRTIO$_3$ (100)

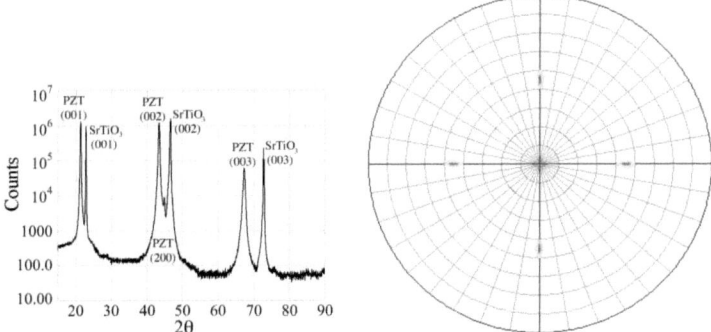

Figure 3.7: (a) Bragg-Brentano $\theta - 2\theta$ scan of a 400nm thick Pb(Zr$_{0.4}$,Ti$_{0.6}$)O$_3$ film on single crystalline SrTiO$_3$ (100) of a 100nm thick Pb(Zr$_{0.4}$,Ti$_{0.6}$)O$_3$ film on SrTiO$_3$ (100). (b) pole figure of the Pb(Zr,Ti)O$_3$ and SrTiO$_3$ [011] directions.

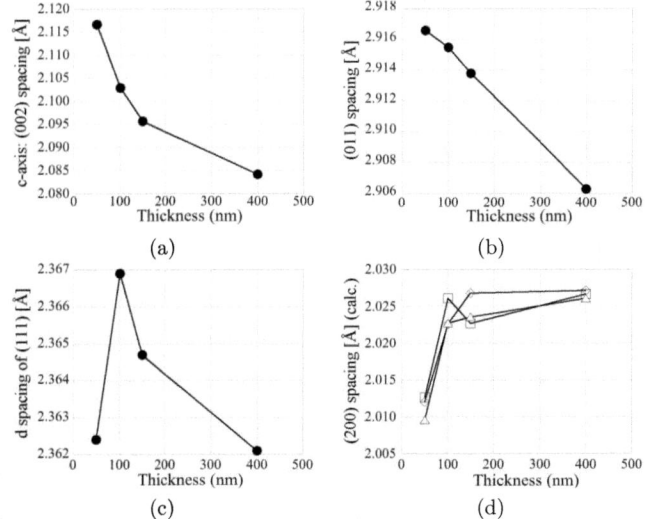

Figure 3.8: X-ray measurements and lattice parameter calculations of Pb(Zr$_{0.4}$,Ti$_{0.6}$)O$_3$ on SrTiO$_3$ (100) for different thicknesses. (a) spacing of (002) planes, (b) spacing of (011) planes, (c) spacing of (111) planes, (d) calculated (200) spacings using the tetragonal hypothesis and the measured values of △ (002) (a) and (111) (c), ◊ (002) (a) and (011) (b), □ (111) (c) and (011) (b).

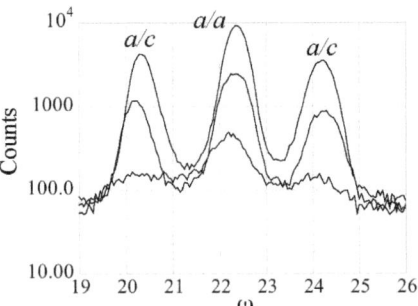

Figure 3.9: Rocking curves of Pb(Zr$_{0.4}$,Ti$_{0.6}$)O$_3$ on SrTiO$_3$ (100) around the peak corresponding to a-domains. a-domains can be seen at $\alpha = 1.8°$ tilted away from the incident [001] direction. This corresponds to the theoretical value using $a = 4.03$Å and $c = 4.16$Å (see Figure 2.10, p.33). The films were 600nm (highest intensity), 400nm and 200nm thick (lowest intensity).

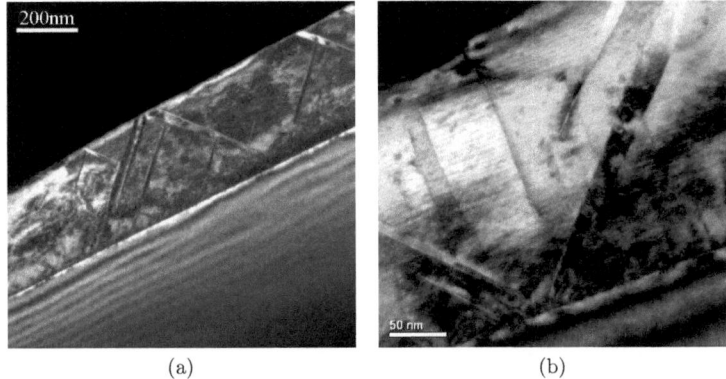

Figure 3.10: Cross-section dark field TEM images showing an epitaxial 400nm thick Pb(Zr$_{0.4}$,Ti$_{0.6}$)O$_3$ film on SrTiO$_3$ (100). (a) domain walls of 90° domains in mostly c-axis oriented material, (b) image showing a/a domains with vertical domain walls.

3.4. PB(ZR,TI)O$_3$ ON SRTIO$_3$ (100)

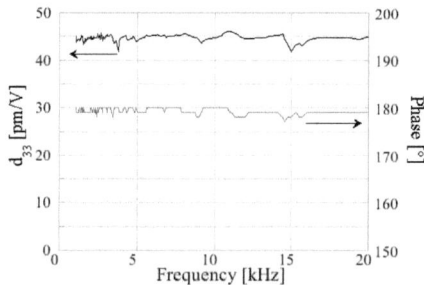

Figure 3.11: Frequency scan of the d_{33} and the phase of a $600nm$ thick Pb(Zr,Ti)O$_3$ film on SrTiO$_3$. The measurement took place on $300 \times 300 \mu m^2$ large Pt top electrodes. The graph shows a stable phase angle at the measurement condition of $17 kHz$ and a d_{33} of $45 pm/V$.

3.4.2 d_{33} Interferometer Measurements on the Continuous Film

The piezoelectric response of the c-axis oriented epitaxial Pb(Zr$_{0.4}$,Ti$_{0.6}$)O$_3$ on SrTiO$_3$ (100) was measured on large scale electrodes using an interferometer set-up. The Nb doped conductive SrTiO$_3$ served as bottom electrode. The top electrode was a $50 \times 50 \mu m$ large Pt area. d_{33} was measured over a frequency range from 1 to $20 kHz$. Only minor resonance frequencies were detected, and above $10 kHz$ the phase decreased slightly (see Figure 3.11, p.53). The measured d_{33} at zero external DC bias was $45 pm/V$. Loop measurements were performed at $17 kHz$ (see Figure 3.12(a), p.54) for three different driving signals: $100 mV$, $1V$, and $3V$ (amplitude). The set-up allows to measure the dielectric constant ϵ_{33} at the same time (b). The loops are characterized by the following features:

- Rectangular shape with abrupt switching, no shift up- or downwards
- Asymmetries in switching behavior
- Spikes just after switching
- Increasing d_{33} at inverse field

In the graph in Figure 3.12(a), p.54, a negative d_{33} corresponds to an up polarization. The rectangular shape of the loop featuring abrupt switching reflects the 180° switching in c-axis oriented ferroelectrics. At zero field, d_{33} was $45 pm/V$, and ϵ_{33} was only 170. The switching behavior is not the same on both sides. Switching from down to the opposite polarization (negative charges on Pt electrode) is less abrupt than in the other case. Not all domains switch immediately as it is the case when switching to the down configuration. This can be related to the different electrode material on both sides, an oxide semiconducting electrode at the bottom (Nb-doped SrTiO$_3$) and a metallic electrode on top. While the polarization can drop into the oxide electrode, this is forbidden on the metallic side. Space charge layers might exist in the vicinity of the Pt electrode to account for the drop of the polarization to zero. As such, domain

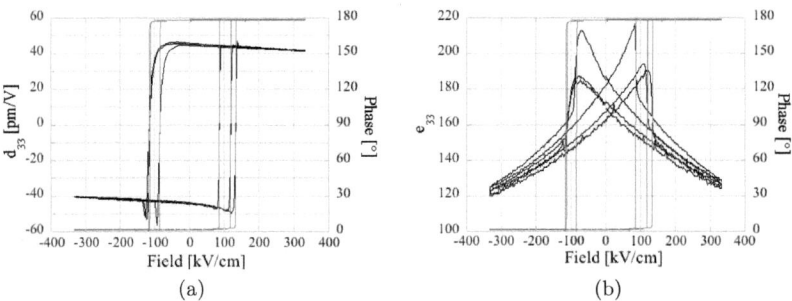

Figure 3.12: d_{33} (a) and ϵ_{33} (b) measurements at $17kHz$ for driving signals of $100mV$, $1V$ and $3V$ (amplitude values). (a) d_{33} (outer loop at $100mV$, center loop at $1V$, inner loop at $3V$), (b) ϵ_{33}. The phase is indicated by a gray line to show when the switching took place.

nucleation will be different on each electrode as well. The observed spikes appear on both sides just after the completion of switching. They are particularly pronounced after switching from down to the opposite direction (negatively charged Pt top electrode). These spikes can not be an artifact from an undefined phase at the switching voltage. They clearly appear just after switching, with a well defined phase. The spikes are related to the switching behavior. Less abrupt switching (left side) leads to more pronounced spikes after switching. Delayed 180° domain switching might lead to these spikes. We also note that the spikes do not appear in the ϵ_{33} plot.

Finally, we observe an increase of d_{33} with inverse field. This is explained considering d_{33} as:

$$d_{33} = 2\epsilon_{33} P_3 Q_{11}$$

where ϵ is the dielectric constant, P the polarization, and Q the electrostrictive constant. ϵ is a function which is inverse proportional to a sum of the uneven order of polarization. If the polarization decreases, ϵ increases. In our case, the polarization loop measurements revealed a decrease of the polarization with the inverse field as seen in Figure 3.13(a), p.55. The faster increase of ϵ compared to the decrease in polarization leads to an overall increase of d_{33}. This is confirmed considering a dielectric constant of 170 at zero field and 190 at switching field Figure 3.12(b) and a polarization of $65pm/V$ at zero field against $60pm/V$ at switching field (Figure 3.13(a), p.55).

3.4.3 Estimations Using the Landau-Ginzburg-Devonshire (LDG) Theory

From the d_{33} and ϵ_{33} measurements on epitaxial (001) Pb(Zr$_{0.4}$,Ti$_{0.6}$)O$_3$ films in Figure 3.12, p.54, it can be seen, that almost exclusively lattice contributions to the response took place. In this case, the LDG theory should well be applicable. First, using the Taylor series development up to the 4th order, the theoretical coercive field can be estimated. Second, using the polarization loop (Figure 3.13(a), p.55), the measurement of the piezoelectric coefficient d_{33} (Figure 3.12(a), p.54) and the dielectric constant ϵ_{33} (Figure 3.12(b), p.54), we use the theory up to the 6th order to calculate the parameters A, B, and C. Knowing the Landau parameters, and using

3.4. PB(ZR,TI)O₃ ON SRTIO₃ (100)

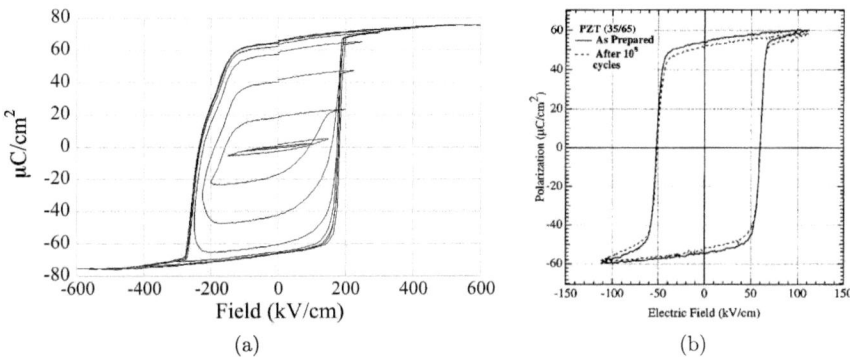

Figure 3.13: Measured negative polarization loops of $600nm$ thick Pb(Zr$_{0.4}$,Ti$_{0.6}$)O$_3$ on SrTiO$_3$ (100) (a), and loops measured by Foster et al.[183] on SrTiO$_3$ (100) with SrRuO$_3$ buffer layer for Pb(Zr$_{0.35}$Ti$_{0.65}$)O$_3$.

the theoretical work from Li et al,[72] we can calculate the critical film thickness below which the polarization would be suppressed at all temperatures.

3.4.3.1 Estimation of the Coercive Field

Including the electrostatic energy created by an external field E and considering volumetric terms up to the 4th order and neglecting surface terms, depolarization as well as gradients in the order parameter, a theoretical value of the coercive field can easily be estimated using the Landau theory:

$$F = \frac{A}{2}P^2 + \frac{B}{4}P^4 - E P$$

The system is always in a minimum state. This is expressed with $\frac{\partial F}{\partial P} = 0$ (with $\frac{\partial^2 F}{\partial P^2} > 0$) and leads to an expression for E (A was set to $A = A_0(T - T_0)$):

$$E = A_0(T - T_0)P + B P^3 \quad (3.1)$$

At zero field, (3.1) can be written as

$$P^2(0) = \frac{A_0}{B}(T_0 - T) \quad (3.2)$$

From the general expression for the polarization $P = \epsilon E$ we deduce that $\frac{\partial E}{\partial P} = \frac{1}{\epsilon}$, and still at zero field

$$\left.\frac{\partial E}{\partial P}\right|_{E=0} = \frac{1}{\epsilon(0)} \quad \text{with (3.1) and (3.2) we write} \quad A_0 = \frac{1}{2\epsilon(0)(T_0 - T)}$$

The thermodynamic coercive field E_c corresponds to the field at which ϵ tends to infinity. This gives

$$\left.\frac{\partial E}{\partial P}\right|_{E_c} = 0 \quad \text{with (3.1)} \quad P^2(E_c) = \frac{A_0(T_0 - T)}{3B} \quad \text{with (3.2)} \quad P^2(E_c) = \frac{1}{3}P^2(0)$$

This allows us to write an expression for E_c

$$E_c = \frac{P(0)}{3\sqrt{3}\epsilon(0)} \quad \text{and} \quad \epsilon(0) = \epsilon_r \cdot \epsilon(0)_{measured} \tag{3.3}$$

The measured values are $P(0) = 65 \mu C/cm^2$ (see Figure 3.13(a), p.55) and $\epsilon(0)_{mesured} = 169$ (see Figure 3.12(b), p.54). ϵ_r is the permittivity of the vacuum ($8.85 \times 10^{-12} F/m$). The calculated coercive field is $E_c = 830 kV/cm$, and the measured one $200 kV/cm$. The theoretical value is much higher than the measured one, but the theory does not take into account the role of nucleation, surface defects and the material of the electrode on the switching. The coercive field can vary considerably as evidenced in the literature, where coercive fields of $50 kV/cm$ (see Figure 3.12(b), p.54) were found for (001) Pb(Zr,Ti)O_3 on SrTiO$_3$ with an SrRuO$_3$ buffer layer. Bellur et al.[41] found $100 kV/cm$ for a $200 nm$ thick (001) PbZr$_{0.53}$Ti$_{0.47}$O$_3$ on Pt on MgO (100) fabricated by sol-gel. The loop in Figure 3.13(a), p.55 also shows the asymmetric characteristic as found in the d_{33} loop.

3.4.3.2 Estimation of the Landau Parameters

Epitaxial films of PZT show a behavior that is close to the one of single crystals / single domains. This means that domain wall contributions to the response are small, giving us the opportunity to test the validity of the LDG theory. To calculate the Landau parameters, we use the following simplified free energy expression (compare with equation on p.15):

$$f = aP^2 + bP^4 + cP^6 \tag{3.4}$$

where f is the free energy $[\frac{J}{m^3}]$. The units of the Landau parameters are: $[a] = \frac{m V}{C}$, $[b] = \frac{m^5 V}{C^3}$, and $[c] = \frac{m^9 V}{C^5}$. a includes the temperature term $(T - T_0)$. In a DC electric field E_{DC} the equilibrium polarization is calculated as real root from the equation:

$$E_{DC} = \frac{\partial f}{\partial P} = 2aP + 4bP^3 + 6cP^5 \tag{3.5}$$

The inverse dielectric constant can be written:

$$\frac{1}{\epsilon} = \frac{\partial^2 f}{\partial P^2} = 2a + 12bP^2 + 30cP^4 \tag{3.6}$$

The small signal piezoelectric response as a function of the DC field is written as:

$$d_{33} = 2 Q_{eff} \, \epsilon(E_{DC}) \, P(E_{DC}) \tag{3.7}$$

For the curve fits, some parameters have been fixed: The polarization at zero field $P(0)$ and the dielectric constant at zero field $\epsilon(0)$. The values have been taken from Figure 3.12(b), p.54 and Figure 3.13(a), p.55:

$$\epsilon(0) = 169$$
$$P(0) = 0.66 C/m^2$$

The expression for $\epsilon(0)$ can be found in equation 3.4. $P(0)$ allows the elimination of one of the unknown parameters. Most convenient is a: From equation 3.5, we obtain at $E_{DC} = 0$:

$$a = -2bP^2(0) - 3cP^4(0) \tag{3.8}$$

3.4. PB(ZR,TI)O$_3$ ON SRTIO$_3$ (100)

The dielectric constant at zero field allows to express the parameter b, using the equations 3.6 and 3.8:

$$\frac{1}{\epsilon_0} = 2a + 12bP^2(0) + 30cP^4(0) = 8bP^2(0) + 24cP^4(0)$$

$$b = \frac{\frac{1}{\epsilon(0)} - 24cP^4(0)}{8P^2(0)}$$

Together with equation 3.8 a can be written as:

$$a = 3cP^4(0) - \frac{1}{4\epsilon(0)} \tag{3.9}$$

Now, there is only c to be adjusted. We tried several ways to find an additional condition for c. In one method we use an additional value of the dielectric constant at high field $\epsilon(h)$ together with equation 3.6. This gives an expression featuring the unknown corresponding polarization $P(h)$. In a first step, $P(h)$ can be estimated and a first value for c be calculated. Knowing now all three Landau parameters, the polarization curve can now be calculated using the measured values of d_{33} and ϵ_{33}, and equation 3.7. Q_{eff} was calculated from the measured values at zero field. At zero field, $P(0) = 0.66 C/m^2$, $\epsilon(0) = 169$, and $d_{33}(0) = 44.2 pm/V$. We find $Q_{eff} = 2.24 \times 10^{-2} m^4/C^2$. The calculated polarization was reintroduced into equation 3.6 to calculate the dielectric constant. The obtained curve was compared with the measured one and $P(h)$ adjusted for the best fit. The following values were found:

$a = -2.12 \times 10^8 mV/C$
$b = 2.96 \times 10^8 m^5 V/C^3$
$c = -7.95 \times 10^7 m^9 V/C^5$

In another method, we calculate $P(h)$ instead of adjusting it for the best fit with respect to the dielectric constant. For this, we consider equation 3.5 and 3.6. From the measurement, the field is $-3.32 \times 10^7 V/m$. Now, we have tow equations and two unknown variables, $P(h)$ and c. Solving the equations yields $P(h) = -0.70205 C/m^2$.

$a = -7.07 \times 10^7 mV/C$
$b = -2.96 \times 10^7 m^5 V/C^3$
$c = 1.69 \times 10^8 m^9 V/C^5$

In this case, a first order transition was found with $b < 0$ and $c > 0$. However, these values are very close to a second order phase transition: A decrease in $P(h)$ of only 0.4% leads to

$a = -8.76 \times 10^7 mV/C$
$b = 1.07 \times 10^7 m^5 V/C^3$
$c = 1.38 \times 10^8 m^9 V/C^5$,

where b is positive, what indicates a second order transition. From equation 3.5 we can calculate the polarization as a function of the field by finding the zero roots. From our calculated Landau parameters, we can calculate the critical film thickness for our Pb(Zr$_{0.4}$,Ti$_{0.6}$)O$_3$ material. We use the equation presented by Li et al.,[72] who used the LDG type shown in this work on p.15:

$$T_c^* = T_0 + \frac{3B^2}{16A_0C} - \left[\frac{2D_{44}}{A_0\delta_1}\left(\frac{1}{a_0} + \frac{1}{b_0}\right) + \frac{2D_{11}}{A_0c_0\delta_3}\right]$$

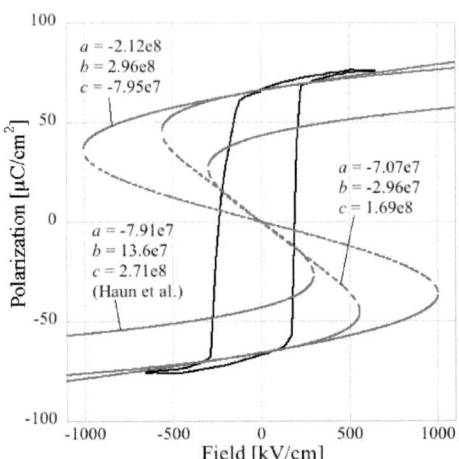

Figure 3.14: Measured polarization loop (black) compared to calculated loops using LDG theory (see p.14). The outer loop was obtained from values which fit best the measured dielectric constant, the center loop from all-calculated values, and the inner loop corresponds to values obtained from bulk Pb(Zr$_{0.4}$,Ti$_{0.6}$)O$_3$ ceramic by Haun et al.[92]

T_c^* is the critical temperature, a_0 and b_0 are the lateral sizes of a feature and are set to infinity for a continuous film. In this case, c_0 becomes the critical film thickness. D was derived from Meyer et al.[271] using their domain wall energy ($D = 4.1 \times 10^{-11} Vm^3/C$) and δ was taken as $1nm$.[72] We calculated the critical film thickness from our values and from those found by Haun et al.[92] (see Figure 3.14, p.58). Considering a critical temperature of $0K$, c_0 is in the range from 0.5 to $1nm$ for all triplets a, b, and c. If the room temperature is taken as the critical temperature, the critical thickness arises to values between 1 and $2nm$. It is interesting to note that the quite different Landau parameters lead to critical thicknesses in the same range. It will be shown later, that we still observed ferroelectricity on $6nm$ thick features.

3.5 Pb(Zr,Ti)O$_3$ on SrTiO$_3$ (111)

Two tetragonal Pb(Zr,Ti)O$_3$ compositions have also been tested for the deposition on SrTiO$_3$ (111): a high Ti containing Pb(Zr$_{0.1}$,Ti$_{0.9}$)O$_3$ and Pb(Zr$_{0.4}$,Ti$_{0.6}$)O$_3$. For the deposition conditions see Table 3.1[3, 4], p.43. The substrate was treated in BHF beforehand. In both cases, only (111) orientations were observed (see Figure 3.15(a), p.59) and pole figure measurements showed epitaxial depositions (Figure 3.15(b), p.59 for Pb(Zr$_{0.4}$,Ti$_{0.6}$)O$_3$ and Figure 3.16, p.61 for Pb(Zr$_{0.1}$,Ti$_{0.9}$)O$_3$). Depositions of Pb(Zr$_{0.4}$,Ti$_{0.6}$)O$_3$ on SrTiO$_3$ (111) showed less surface roughness than high Ti containing Pb(Zr,Ti)O$_3$. AFM measurement on $400nm$ thick Pb(Zr$_{0.4}$,Ti$_{0.6}$)O$_3$ revealed a roughness of 7Å and a $200nm$ thick Pb(Zr,Ti)O$_3$ film with 90%

3.5. PB(ZR,TI)O$_3$ ON SRTIO$_3$ (111)

Ti was twice as rough and shoed a triangular surface pattern (see Figure 3.17, p.61). XRD measurements of the d-spacings of different orientations showed that the Pb(Zr$_{0.4}$,Ti$_{0.6}$)O$_3$ film crystallized in the rhombohedral phase. No splitting neither between the (100) and (001) planes and between the (110) and (011) planes could be observed, but the space diagonals [111] and [11$\bar{1}$] showed different values (see Table 3.3, p.60). The highly tetragonal Pb(Zr,Ti)O$_3$ composition Ti (90%) on SrTiO$_3$ (111) showed tetragonal but also rhombohedral features (see Table 3.3, p.60). The rhombohedral feature was the observed different length of the [111] (2.351Å) and [11$\bar{1}$] (2.318Å) direction (see scans in Figure 3.18(a), p.62). The tetragonal features were the observed splitting between the [200] and [002] and between the [220] and [022] directions (see scans in Figure 3.18(b–c), p.62). Large FWHM values of the rocking curves of more than 1° were found in the case of Pb(Zr,Ti)O$_3$ containing 90% Ti (see Figure 3.19, p.62 for scans, Table 3.3, p.60 for values). FWHM values for the single crystalline SrTiO$_3$ substrate are 0.2°. Very large FWHM values were found for the [022] direction (2.6°) showing a triangular shape.

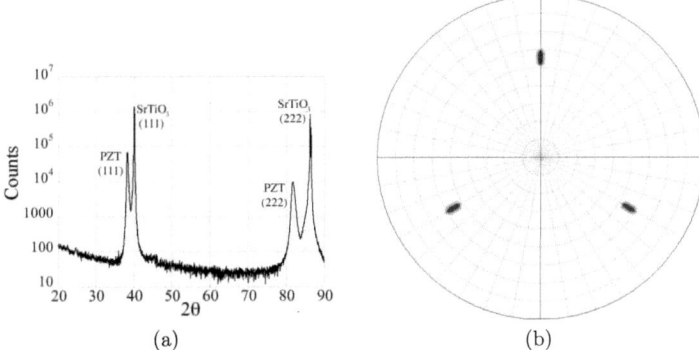

Figure 3.15: (a) X-ray Bragg-Brentano $\theta - 2\theta$ scan of a 200nm thick Pb(Zr$_{0.4}$,Ti$_{0.6}$)O$_3$ film on SrTiO$_3$ (111). (b) pole figure of the [001] directions of Pb(Zr$_{0.4}$,Ti$_{0.6}$)O$_3$ (dark, $2\theta = 21°.67$) and SrTiO$_3$ (gray, $2\theta = 22°.80$). No tetragonal distortions have been observed for this tetragonal PZT on SrTiO$_3$ (111).

Substrate	SrTiO$_3$ (111)		SrTiO$_3$ (111)		SrTiO$_3$ (001)	
Composition	Pb(Zr$_{0.1}$,Ti$_{0.9}$)O$_3$		Pb(Zr$_{0.4}$,Ti$_{0.6}$)O$_3$		Pb(Zr$_{0.4}$,Ti$_{0.6}$)O$_3$	
	d [Å]	Rocking FWHM [°]	d [Å]	Rocking FWHM [°]	d [Å]	Rocking FWHM [°]
(111)	2.351	1.690	2.370	0.787	2.365	0.601
(11$\bar{1}$)	2.318	1.872	2.345	0.804	-	-
(200)	1.992	1.320	2.0384	0.656	2.026	0.459
(002)	2.073	2.128	-	-	2.084	-
(220)	1.412	0.953	1.4465	0.794	1.412	0.565
(022)	1.442	2.593	-	-	1.442	-

Table 3.3: XRD data of tetragonal Pb(Zr,Ti)O$_3$ compositions on SrTiO$_3$ (111) compared with Pb(Zr,Ti)O$_3$ on SrTiO$_3$ (100). Pb(Zr$_{0.4}$,Ti$_{0.6}$)O$_3$ on SrTiO$_3$ (111) crystallizes in the rhombohedral phase (different length of d_{111} and $d_{11\bar{1}}$). Pb(Zr$_{0.1}$,Ti$_{0.9}$)O$_3$ on SrTiO$_3$ (111) shows rhombohedral and tetragonal features. Pb(Zr$_{0.4}$,Ti$_{0.6}$)O$_3$ on SrTiO$_3$ (100) is purely tetragonal. Typical rocking curve FWHM values for SrTiO$_3$ are 0.20° for (100) and 0.25° for (111) substrates.

3.5. PB(ZR,TI)O$_3$ ON SRTIO$_3$ (111)

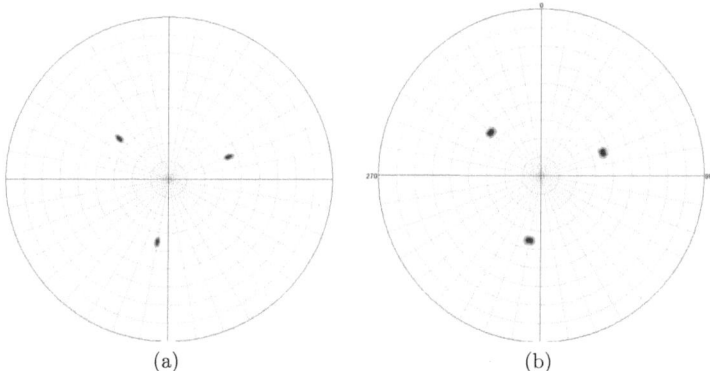

Figure 3.16: XRD pole-figure measurement of [220] directions of Pb(Zr$_{0.1}$,Ti$_{0.9}$)O$_3$ on SrTiO$_3$ (111). (a) SrTiO$_3$ (2θ = 66°.07), (b) Pb(Zr,Ti)O$_3$ (2θ = 64°.56). The spots are located at χ = 36° with respect to the surface (111) normal. The tetragonal distortions of the PZT film are evidenced by a splitting of the peaks in ϕ.

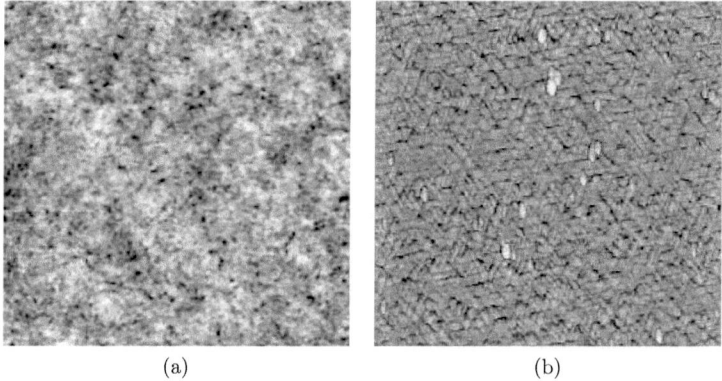

Figure 3.17: 5 × 5μm^2 AFM topographies of tetragonal Pb(Zr,Ti)O$_3$ (111) surfaces on SrTiO$_3$ (111) for two different compositions: (a) Pb(Zr$_{0.4}$,Ti$_{0.6}$)O$_3$ 400nm, (b) Pb(Zr$_{0.1}$,Ti$_{0.9}$)O$_3$ 200nm. The triangular structure can clearly be seen in (b). The root mean square roughness in (a) was 7Å, in (b) 14Å.

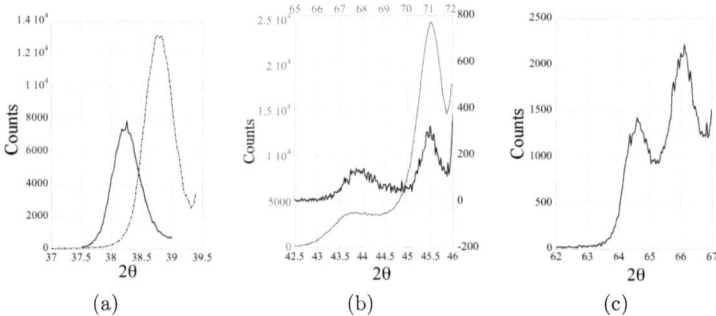

Figure 3.18: $\theta - 2\theta$ XRD measurements of Pb(Zr$_{0.1}$,Ti$_{0.9}$)O$_3$ on SrTiO$_3$ (111) showing the splitting of the [111], [110] and [100] directions. (a) plain line: [11$\bar{1}$] measured at $\chi = 0°$, dashed line: [111] measured at $\chi = 71°$. (b) [200] (gray line, 2θ-axis on top) and [300] (dark line, 2θ-axis at bottom) directions measured at $\chi = 55°$. (c) [220] direction measured at $\chi = 35°$. The values for the d-spacings can be found in Table 3.3, p.60 .

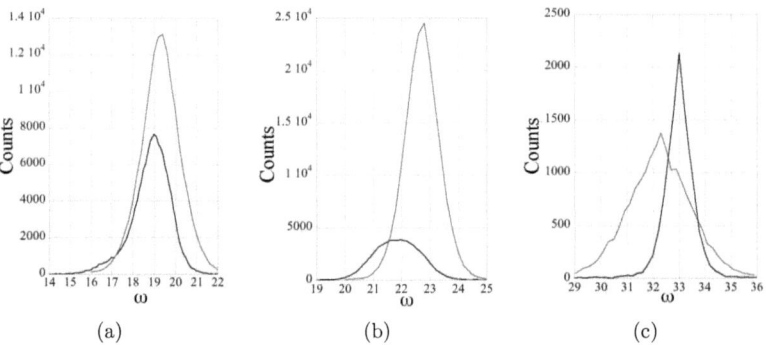

Figure 3.19: Rocking curves of Pb(Zr$_{0.1}$,Ti$_{0.9}$)O$_3$ on SrTiO$_3$ (111). (a) [111] (dark) and [11$\bar{1}$] (gray), (b) [002] (dark) and [200] (gray), (c) [220] (dark) and [022] (gray).

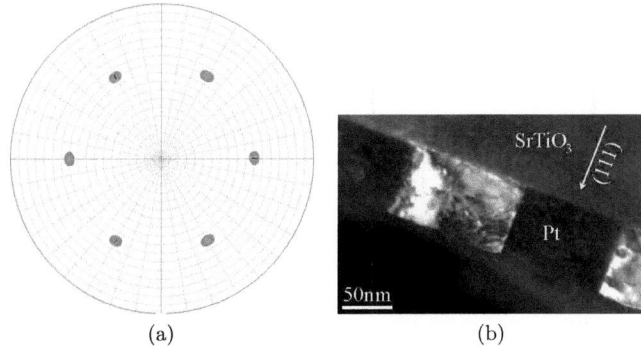

Figure 3.20: (a) Pole-figure of Pt (200) directions measured at $2\theta = 46°.35$ ($d_{Pt(100)} = 3.914\text{Å}$), (b) dark field TEM image showing one type of twins in Pt.

3.6 Pt on SrTiO$_3$ (111)

In the additive route, we used epitaxial Pt on SrTiO$_3$ (111). The substrates were treated in BHF before deposition. For the Pt deposition, it was found that higher growth rates and lower temperature favored smoother films. The deposition condition used for these films is found in Table 3.1[7], p.43. Figure 3.20(a), p.63 shows the pole figure of Pt (200) planes on SrTiO$_3$ (111). The spots are situated at $\chi = 55°$ proving the (111) growth of Pt on SrTiO$_3$. Due to the close lattice parameter of Pt and SrTiO$_3$ ($d_{Pt_3,(100)} = 3.914\text{Å}$, $d_{SrTiO_3,(100)} = 3.900\text{Å}$) three intense (200) spots from SrTiO$_3$ also appear in the figure, which are centered on the gray Pt (200) spots. Six Pt spots instead of only three can be seen. There are two solutions for Pt to fit the SrTiO$_3$ (111) planes. One takes over the crystalline (111) orientation of SrTiO$_3$ and the other is rotated by 60° with respect to SrTiO$_3$. This corresponds to a stacking fault in the atomic arrangement in the [111] direction leading to twinning of the Pt. Figure 3.20(b), p.63 highlights the twinned structure. It shows a dark field TEM image of a $70nm$ thick Pt film where one type of Pt-twin was taken into consideration, which appears in a bright contrast. AFM measurements revealed very low surface roughness of less than 4Å (root mean square) for a $70nm$ thick Pt film. A low surface roughness is important for subsequent Pb(Zr,Ti)O$_3$ deposition, because such a surface provides less nucleation sites for PZT.

3.7 Pt on MgO (100) and SrTiO$_3$ (100) with Ir Seeding Layer

Direct growth of epitaxial Pt on MgO (100) and SrTiO$_3$ was difficult to obtain as Pt likes to grow in the [111] direction (see p.28). In other works and in our experiments, it was shown that Pt (100) nucleates on MgO (100) only at low pressures, high temperatures and very low growth rates. Table 3.5, p.66 shows the results for Pt depositions on MgO (100) single crystalline substrates at different conditions. The degree of orientation was evaluated by $\theta - 2\theta$ XRD measurements (see p.44), comparing the intensity arising from the Pt (111) with the MgO (200) planes. To decrease the deopsition rate to a minimum, RF power was used and the substrate

was rotated at $10 rpm$. From this table, it can be seen, that pure Pt (100) growth was obtained for growth ratres below $1nm/min$ and at a nominal substrate temperature of $700°C$.
We found that the initial nucleation of Pt was important. Once a thin Pt seeding layer of $10nm$ was deposited on MgO, the epitaxial growth of Pt at higher growth rates was possible. Therefore, not only the lowest surface energy of the Pt (111) atomic planes plays an important role, but also the interface energy between Pt and MgO. After the deposition of a thin epitaxial Pt seed layer on MgO (100), a second Pt film deposited at higher growth rates (5.2nm/min) was still epitaxial. In XRD, the Pt (111) appeared again at a deposition rate of $8.5nm/min$. The Pt seeding layer is the crucial point but the deposition takes place at conditions which are close to the limit of the sputtering facility ($800°C$, $2mT$).
The same depositions on SrTiO$_3$ (100) were never successful. The XRD measurements always showed a Pt (111) peak. AFM images of Pt depositions on MgO and SrTiO$_3$ (100) are shown in Figure 3.21, p.65. The typical rectangular features of a (100) epitaxy is observed in both cases. On SrTiO$_3$ (100) the Pt forms uncontrollable outgrows which lead to a rough surface, and which are probably responsible for the Pt (111) peak in X-ray diffraction.
We found that Ir is much easier to grow in the [100] direction. Iridium is a face-centered cubic metal of the platinum group.[272] Ir is not as inert as Pt and oxidizes during the subsequent growth of Pb(Zr,Ti)O$_3$. We used an Ir seeding layer and subsequent Pt growth on Ir in order to achieve an epitaxial and inert Pt (100) surface for the subsequent deposition of PZT.
It is generally found, that the surface energy of fcc metals is lowest for (111) planes, followed by (100) and (110).[143] This is also true for Ir.[273] From this point of view, (111) growth would be expected. Interface energies and lattice match play an important role for epitaxial growth. If no chemical bonds were created between two interface, the interface energy is just the addition of the two surface energies. Nevertheless this case is an ideal one and the interface energy has to be into taken account as a function of bond strength between two surfaces.[171] Pt has a very low affinity to oxygen[158] in contrast to Ir of which the oxide IrO$_2$ is used a transparent conductive layer in many applications. Therefore, Ir undergoes much stronger bonds with the topmost layer of MgO than Pt. Hence, Ir grows easily in the [100] direction even though the (111) plane has a lower surface energy. The chemical affinity of Ir is the dominating factor of it's epitaxial growth on MgO even though the misfit between MgO and Ir is very large (8.5% at the growth temperature of $700°C$, see Table 3.1, p.43).
Iridium was epitaxially deposited on MgO at a nominal temperature of $700°C$ ($600°C$ on the surface) and at a high growth rate of $8nm/min$ (see Table 3.1[5], p.43). Figure 3.22(a–c), p.67 shows AFM images and a pole-figures for $85nm$ thick Ir on MgO (100). Dense dislocation formation is expected to happen considering the large lattice misfit between Ir and MgO (see Table 3.4, p.65), which leads to observable surface patterns. This was also observed in the case of Pb(Zr,Ti)O$_3$ depositions on SrTiO$_3$ (111) and (100) (seeFigure 3.17(b), p.61 and Figure 3.6, p.49). AFM images of Ir depositions on MgO (100) showed needle-like surface patterns. The needles point in a direction which is at $45°$ with respect to the [100] directions. The surface root means square roughness was low (4Å). Pole-figure measurements of the Ir [111] directions (Figure 3.22(c), p.67) proved the epitaxial relationship with MgO (b). Subsequent Pt depositions on Ir were performed at $700°C$ at a deposition rate of $2.3nm/min$. Pt grew epitaxially on Ir, as evidenced by pole-figure measurements (e). With a $35nm$ thick Pt layer on top of Ir, the surface roughness measured by an AFM was increased only to 6.4Å. The Pt deposition led to a terrace-like growth onto the Ir needles (d).

3.8. TIO₂ ON PT (100) AND PT (111)   65

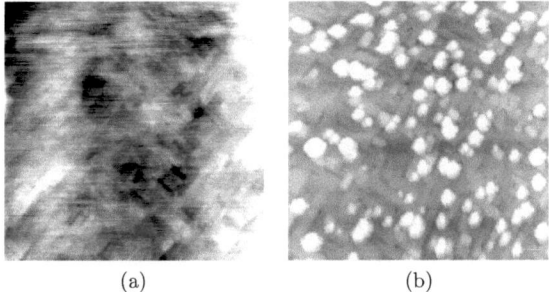

(a)   (b)

Figure 3.21: $5 \times 5\mu m^2$ AFM scans on Pt surfaces deposited on MgO (100) (a) and SrTiO$_3$ (100) (b). The $50nm$ thick Pt was directly deposited on the substrates. The deposition included a $10nm$ thick Pt seed layer as discussed in the text (see Table 3.5, p.66). The roughness in (a) was 4Å, and in (b) $4nm$.

3.8 TiO$_2$ on Pt (100) and Pt (111)

As we use TiO$_2$ as seeding layer for Pb(Zr,Ti)O$_3$ growth, it is of particular interest to know how the seed layer grows on Pt (100) and Pt (111). Given the weak response of TiO$_2$ in X-ray diffraction we had to deposit a thick layer of TiO$_2$ of $200nm$ to establish pole-figures. The deposition conditions for TiO$_2$ are shown in Table 3.1[1], p.43. An excellent review about TiO$_2$ was recently published.[275] TiO$_2$ is interesting for catalytic reactions, and most studies in literature have been carried out on single crystalline rutile TiO$_2$ (110) surfaces.[275] TiO$_2$ (110) is thermodynamically the most stable facet and most studies focus on this surface.[165,276] The reactivity of TiO$_2$ surfaces is well known,[277,278] and surface defects play an important role in catalyzed photoelectrolysis of water.[278] The presence of unstructured dangling bonds on the unreconstructed (001) surface renders this surface thermodynamically unstable and their presence is thus responsible for the experimentally observed, thermally induced reconstruction or facetting of this surface[276] which can be oxygen deficient.[277] Pt has been deposited onto TiO$_2$ (100) surfaces,[279] however, the deposition of epitaxial TiO$_2$ on Pt surfaces has not been reported yet. Only the homoepitaxial growth of TiO$_2$ (100) was reported.[280] Surface free energies at the liquid state[281] were experimentally evaluated at the melting temperature. For TiO$_2$ this energy was $0.02eV/Å^2$. For PbO the value is about 3 times lower. This is an indication that PbO should wet well TiO$_2$. The value for Pt is about 6 times higher with respect to TiO$_2$.[281] Pt (100) has a calculated surface energy of $1.126eV/atom$ $(0.146eV/Å^2)$.[167] Therefore TiO$_2$ wets

	d_{100} [Å]	exp $[10^6 K^{-1}]$
MgO	4.2004	14.5[218]
Ir	3.8600	6.4[274]
Pt	3.9340	8.8[274]

(a)

	Misfit δ [%]
MgO-Ir	8.5
Ir-Pt	2.0

(b)

Table 3.4: (a) Lattice parameters and (b) thermal expansion coefficient for Pt, Ir and MgO, and the calculated misfit[151] at $600°C$ with $\delta = \frac{d_{substrate} - d_{film}}{d_{Substrate}}$.

well other surfaces due to its low surface energy, but high surface energy materials with low oxygen affinity tend to cluster on the rutile substrate.[241, 275]
Our depositions were carried out at 650°C and TiO_2 crystallized in the rutile phase. The crystal structure is depicted in Figure 3.23(a), p.68. TiO_2 grew epitaxially on Pt (111) in the [100] direction, even though the lowest energy plane in TiO_2 is (110). An epitaxial effect between TiO_2 (100) and Pt (111) is responsible for this. Figure 3.24, p.68 shows the pole-figure of the [110] direction of Pt (a) and TiO_2 (b). In the case of TiO_2 (100) the peaks are situated at $\chi = 45°$ which proves TiO_2 (100) growth. The twinning of the Pt is translated into the orientation of TiO_2: It also shows 6 peaks. This can be attributed to epitaxial deposition on single Pt twins or to twinning of TiO_2 itself. Considering just one Pt layer, a 6-fold-axis can be defined that accounts for the observed 60° rotation. Therefore, two TiO_2 (100) growth domains can be found. Figure 3.26(a), p.69 shows the two solutions by which the rectangular base of (100) TiO_2 can be matched with the hexagonal symmetry of Pt (100). On a Pt (111) surface, adjacent Pt atoms in the [110] direction are spaced by 2.766Å. This corresponds well to half the length of the diagonal in a TiO_2 (100) plane (5.463Å) and to the a-axis of TiO_2 (2.958Å). On Pt (100), TiO_2 also grew in the rutile phase. Again, TiO_2 grew in the [100] direction. This was evidenced by XRD pole-figure measurements where the TiO_2 (110) peaks were located at $\chi = 45°$ (see Figure 3.25(b), p.69), and by TEM diffraction. 3 growth domains exist which are repeated with respect to a 4-fold-axis leading to a total of 12 peaks. One preferred growth domain has a much higher intensity than the others. The two lower intense peaks are situated at $\phi = 28°.5$ on the left and right side of a intense peak. The high intense TiO_2 (110) peak is rotated by 45° with respect to the Pt (110). The atomic arrangement of the three growth domains are depicted in Figure 3.26(b), p.69. Corresponding to the pole-figure in Figure 3.25(b), p.69, the a and c-axis of TiO_2 should match the atomic arrangement of Pt in the [110] direction. The smallest mismatch is found for the growth with the highest intensity in XRD. it is the Pt [110] spacing of adjacent atoms (2.766Å) which matches with the c-axis of TiO_2 (2.958Å). In the direction perpendicular, the a-axis of TiO_2 (4.593Å) does not match with the [110] spacing of Pt, but a good match is obtained after aligning 3 unit cells of TiO_2 in the [100] direction on 5 units cells of Pt in the [110] direction. In this case, the mismatch is only 0.4%. The AFM scan on $200nm$ thick TiO_2 showed rectangular features as seen in Figure 3.25(c), p.69. The root mean square roughness was $5.3nm$. The dielectric constant was measured as exactly 100.

#	T [°C]	rate [nm/s]	$P_{DC}[W]$	$P_{RF}[W]$	$P^*_{RF_s}[W]$	$\frac{I_{Pt,(002)}}{I_{MgO,(002)}}$	$\frac{I_{Pt,(111)}}{I_{MgO,(002)}}$
[1]	RT	2	25	-	0	0	1.2
[2]	RT	1.3	25	-	25	0	1.2
[3]	700	1.3	25	-	25	0.5	1.1
[4]	700	0.34	-	30	0	0.9	0
[5]	550	0.34	-	30	0	0.9	0.3
[6]	700	1	-	60	0	1.2	0

Table 3.5: Deposition conditions for the direct deposition of Pt on MgO (100) and presence of the X-ray peak of Pt (111) and Pt (200). All depositions were carried out at $3mT$ and an Ar flow of $10sccm$. The substrate was rotated at $10rpm$. DC as well as RF power was used. An RF power $P^*_{RF_s}$ was applied to the substrate to further reduce the deposition rate. Low deposition rates (0.34 to $1nm/min$) and high temperatures (700°C) were necessary to obtain pure Pt (100) growth.

3.8. TIO$_2$ ON PT (100) AND PT (111)

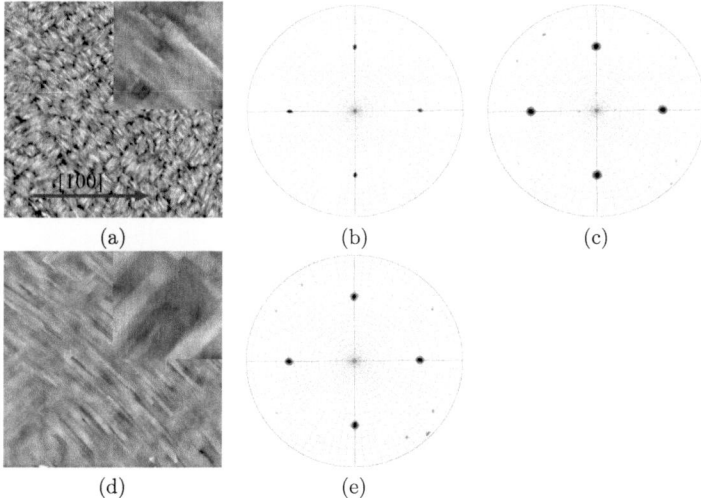

Figure 3.22: Epitaxial depositions of $85nm$ Ir on MgO (100) and subsequent deposition of $35nm$ Pt on Ir. After $85nm$ Ir deposition (a) $5 \times 5\mu m^2$ (inset: $500 \times 500 nm^2$) AFM image with indicated [100] direction, (b) pole-figure of the MgO [111] direction, (c) pole-figure of the Ir [111] direction. After subsequent $35nm$ Pt deposition on Ir: (d) $5 \times 5\mu m^2$ (inset: $500 \times 500 nm^2$) AFM image on Pt, (e) pole-figure of the Pt [111] direction. The root mean square value of the Ir surface roughness as measured by AFM was 4Å and on Pt 6.4Å.

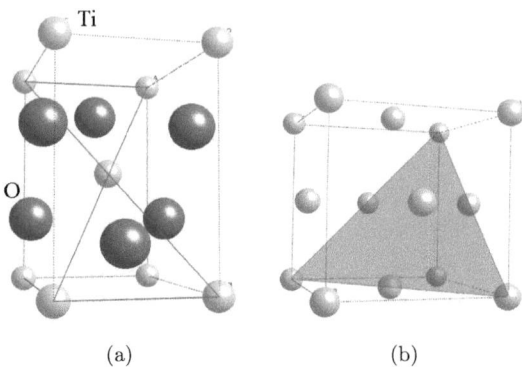

Figure 3.23: Crystal structure of (100) rutile TiO$_2$ and Pt. (a) TiO$_2$ rutile with indicated (110) plane as measured by pole-figure measurements (see Figure 3.24, p.68 and Figure 3.25, p.69). Gray: Ti, Dark: O. The structure is tetragonal with $a = 4.5933$Å and $c = 2.9592$Å. The position of the O atom is $x = y = 0.3$. (b) Pt with indicated (111) plane.

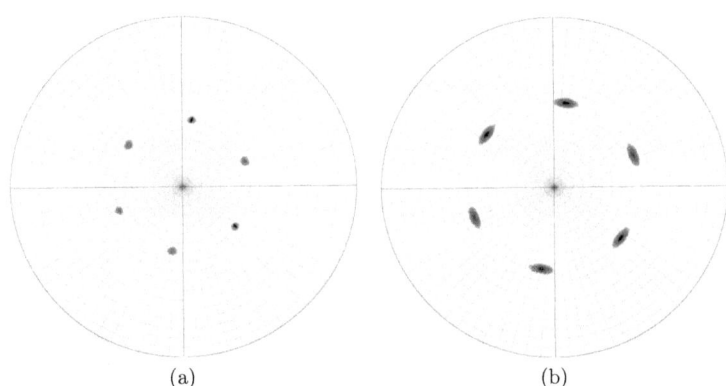

Figure 3.24: Pole-figures of $200nm$ TiO$_2$ on epitaxial Pt/SrTiO$_3$ (111). (a) Pt (111) growth as evidenced by the Pt (220) peaks measured at $2\theta = 67°.484$, located at $\chi = 35°$. (b) TiO$_2$ (200) growth on Pt (111) as evidenced by the TiO$_2$ (110) peaks measured at $2\theta = 27°.446$, located at $\chi = 45°$.

3.8. TIO₂ ON PT (100) AND PT (111)

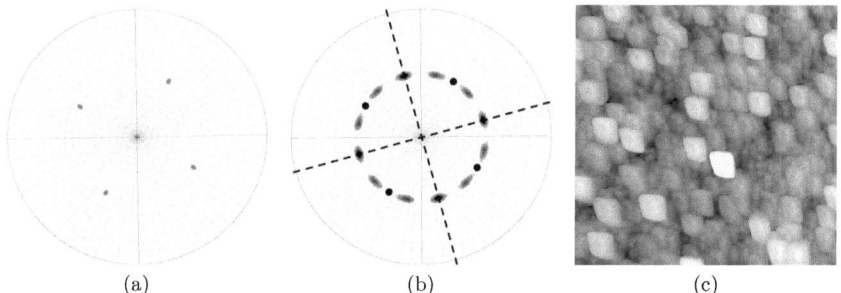

Figure 3.25: Pole-figures of $200nm$ TiO₂ on epitaxial Pt/Ir/MgO (100). (a) Pt and MgO (220) peaks measured at $2\theta = 67°.420$. (b) TiO₂ (110) peaks measured at $2\theta = 27°.440$. Superimposed are the black spots which indicate the (220) reflection of the substrate, and dashed lines running through the most intense TiO₂ (110) reflections. These lines are situated at 45° with respect to the (220) reflections from the substrate. (c) $2.5 \times 2.5 \mu m^2$ AFM scan of the $200nm$ thick TiO₂.

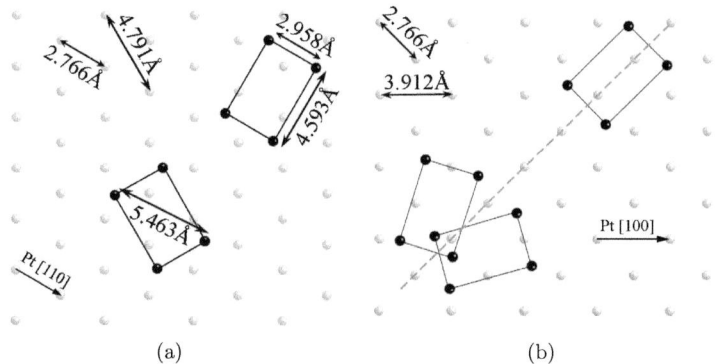

Figure 3.26: (a) TiO₂ on Pt/SrTiO₃ (111) and Pt/Ir/MgO (100) (b). The dark atoms are Ti atoms corresponding to the (100) plane (Figure 3.23, p.68). The position of the TiO₂ (100) plane with respect to Pt (111) (a) and Pt (100) (b) are deduced from the [110] measurements of the pole-figures in Figure 3.24, p.68 and Figure 3.25, p.69. The dashed line in (b) corresponds to the projectino of the TiO₂ [110] direction on the (100) plane. The projection of the Pt [110] on the same plane has the same direction as the indicated Pt [110]. This corresponds to the high intense XRD spots in Figure 3.25(b), p.69. The two other growth domains are rotated by ±28°.5 with respect to the first one.

(a) (b)

Figure 3.27: SEM images of $100nm$ Pb(Zr$_{0.4}$,Ti$_{0.6}$)O$_3$ on single crystal Pt (100). (a) zone of dense Pb(Zr,Ti)O$_3$ nucleation, (b) zone of weak nucleation near substrate border.

3.9 Pb(Zr,Ti)O$_3$ Depositions

In the additive route, the nucleation-behavior of PZT on Pt is of great importance. In this route, we use Pt (111) and Pt (100) substrates. It is already well known, that direct depositions of high Zr containing Pb(Zr,Ti)O$_3$ films on Pt (111) lead to mixed orientations and to the nucleation of pyrochlore, and that PbTiO$_3$ depositions lead to mixed a and c-axis oriented growth. We investigated the direct growth of Pb(Zr$_{0.4}$,Ti$_{0.6}$)O$_3$ and PbTiO$_3$ on the Pt (100) surface. The effect of a $2nm$ thick TiO$_2$ seed layer on the Pb(Zr$_{0.4}$,Ti$_{0.6}$)O$_3$ orientation on Pt (100) and (111) was investigated as well.

3.9.1 Direct Deposition of Pb(Zr,Ti)O$_3$ on Pt (100)

Pt is not an ideal substrate for Pb(Zr,Ti)O$_3$ nucleation. To nucleate Pb(Zr,Ti)O$_3$ directly on Pt high lead oxide or titania fluxes are required. This is particularly the case on the lowest energy Pt (111) surface where direct deposition of Pb(Zr,Ti)O$_3$ leads to pyrochlore phase with some traces of perovskite.[282] On Pt (100), however, the nucleation is facilitated by the higher surface energy of this plane. This was observed for Pb(Zr,Ti)O$_3$ depositions on a (100) single crystal of Pt (see Table 3.1[3], p.43 for deposition condition). Figure 3.27(a), p.70 shows a region of denser Pb(Zr,Ti)O$_3$ nucleation on Pt (100). The nominal thickness was $100nm$. Island growth is observed leading to a discontinuous film even for a film of that thickness. Mainly cubic shaped crystals showing a uniform orientation suggests epitaxial PZT. Triangular (111) crystals also exist. Between the crystals, an underlying surface layer can be seen. (b) shows an SEM image of a region at the border of the substrate where the nucleation density was greatly decreased. Here, the underlying film can clearly be seen.

3.9.2 PbTiO$_3$ on Pt (100)

Epitaxial c-axis oriented growth of a $200nm$ thick PbTiO$_3$ film (for deposition conditions see Table 3.1[2], p.43) was obtained on Pt (100)/Ir/MgO as evidenced by (111) pole-figure measurements in Figure 3.28(a–c), p.72. The lead flux was increased by increasing the power fom 150W to 200W. A PbO termination stabilizes the (001) growth due to the layered structure of PbO. In the pole-figures, all spots are located at $\chi = 55°$ which indicates (001) growth. The PbTiO$_3$ spots (c) are even narrower than the ones of Pt, which indicates excellent orientation. A $2.5\mu m \times 2.5\mu m$ AFM scans was acquired on another sample for a deposition of $5nm$ PbTiO$_3$ on Pt (100) (d). A non-continuous film with high roughness (root mean square $1.7nm$) was observed at that deposition stage. The AFM image of the $200nm$ thick film is shown in (e). The roughness was $6.3nm$. For comparison, if $5nm$ of PbTiO$_3$ was deposited on Pt (100)/Ir/MgO (100) with a $2nm$ thick TiO$_2$ seed layer, dense nucleation of a continuous film was observed with AFM scans (f).

3.9.3 Pb(Zr,Ti)O$_3$ on Pt (100) with TiO$_2$ Seed- and PbTiO$_3$ Starting Layer

Direct deposition of Pb(Zr$_{0.4}$,Ti$_{0.6}$)O$_3$ on Pt (100) with TiO$_2$ seed layer led to pure (111) growth. To investigate whether it is possible to switch from (111) to (100) growth, a $10nm$ thick PbTiO$_3$ starting layer was applied on the TiO$_2$ seed layer prior to Pb(Zr,Ti)O$_3$ deposition. It was found that only few (001) oriented material was obtained in that way. Figure 3.29(b), p.73 shows the (110) pole-figure of $200nm$ Pb(Zr$_{0.4}$,Ti$_{0.6}$)O$_3$ deposited on $10nm$ PbTiO$_3$/$2nm$ TiO$_2$/Pt (100)/Ir/MgO (100). The orientation of Pb(Zr,Ti)O$_3$ is dictated by the underlying TiO$_2$ (compare with TiO$_2$ depositions on Pt (100) in Figure 3.25(b), p.69). The 12 main peaks of Pb(Zr,Ti)O$_3$are located at $\chi = 35°$ which corresponds to the (111) growth. The deposition of $10nm$ PbTiO$_3$ could only switch a small part of the material. This led to 4 spots of weak intense Pb(Zr,Ti)O$_3$(001) at $\chi = 45°$, superimposed with the (220) spots from Pt (a), indicating a (001) epitaxial growth on Pt (100). Applying higher PbO flux might switch even more material from (111) to (001) growth. Hence, TiO$_2$ is a very efficient seed layer for (111) growth of PZT.

3.9.4 Pb(Zr,Ti)O$_3$ on Pt (111) with TiO$_2$ Seed Layer

Pb(Zr$_{0.4}$,Ti$_{0.6}$)O$_3$ was epitaxially grown on $2nm$ TiO$_2$/Pt/SrTiO$_3$ (111). Figure 3.30(a), p.73 shows the (110) pole-figure of $200nm$ thick Pb(Zr,Ti)O$_3$. Pb(Zr,Ti)O$_3$ shows the same in-plane orientation as the twinned Pt (compare with Figure 3.20, p.63), exhibiting 6 peaks instead of 3 as in the case of SrTiO$_3$ (Figure 3.30(b), p.73). A sample with $300nm$ Pb(Zr$_{0.4}$,Ti$_{0.6}$)O$_3$ on Pt (111) covered by $2nm$ TiO$_2$ was investigated by TEM (Figure 3.31, p.74). The image shows the Pb(Zr,Ti)O$_3$ film on the left, followed by the (111) Pt and the single crystalline SrTiO$_3$ (111) substrate. The two types of domains existing in (111) oriented films are observed. A very dense domain structure can be seen in the top region of the film with domain widths of less than $15nm$. The observed inclined domains constitute a first type. The observed vertical domain wall is the separation between domains of type 2.

CHAPTER 3. GROWTH OF EPITAXIAL THIN FILMS

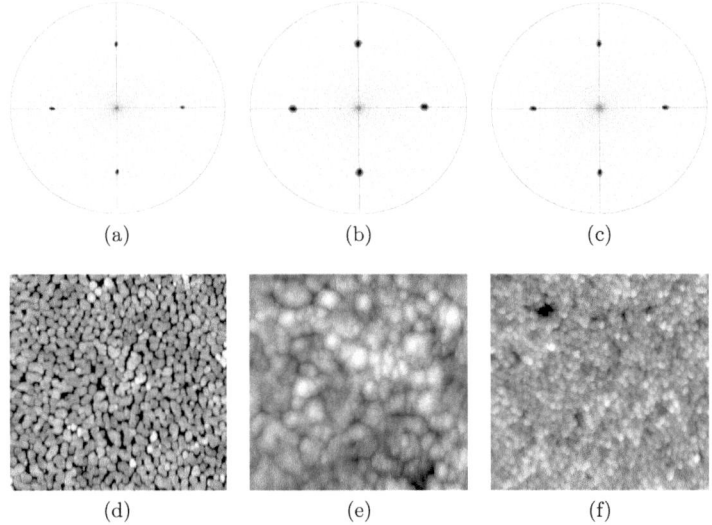

Figure 3.28: (111) pole-figure of $200nm$ PbTiO$_3$ on Pt/Ir/MgO (100). (a) MgO (111) measured at $2\theta = 36°.872$, (b) Pt (111) measured at $2\theta = 39°.726$, (c) PbTiO$_3$ (111) measured at $2\theta = 38°.270$. $2.5 \times 2.5\mu m^2$ AFM scans of topographies obtained on depositions on PbTiO$_3$/Pt/Ir/MgO (100). (d) $5nm$ PbTiO$_3$ on Pt, (e) $200nm$ PbTiO$_3$ on Pt, (f) $5nm$ PbTiO$_3$ on $2nm$ TiO$_2$ on Pt (100). RMS roughness: (d) $1.7nm$, (e) $6.3nm$, (f) 1.2nm.

3.9. PB(ZR,TI)O₃ DEPOSITIONS

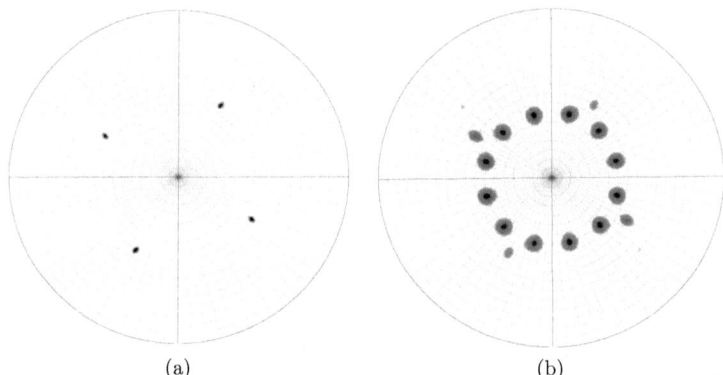

(a)   (b)

Figure 3.29: Pole-figures of $200nm$ Pb(Zr$_{0.4}$,Ti$_{0.6}$)O$_3$ on Pt (100)/Ir/MgO with $2nm$ TiO$_2$ seed layer. Prior to the Pb(Zr$_{0.4}$,Ti$_{0.6}$)O$_3$ deposition a $10nm$ thick PbTiO$_3$ starting layer was applied to enhance nucleation. (a) Pole-figure of (220) Pt measured at $2\theta = 67°.487$. The 4 peaks are located at $45°$ which corresponds to the Pt (100) growth on Ir/MgO (100) as discussed before. (b) Pb(Zr$_{0.4}$,Ti$_{0.6}$)O$_3$ (110) measured at $2\theta = 30°.858$. The peaks are located at $35°$ which indicates PZT (111) orientation.

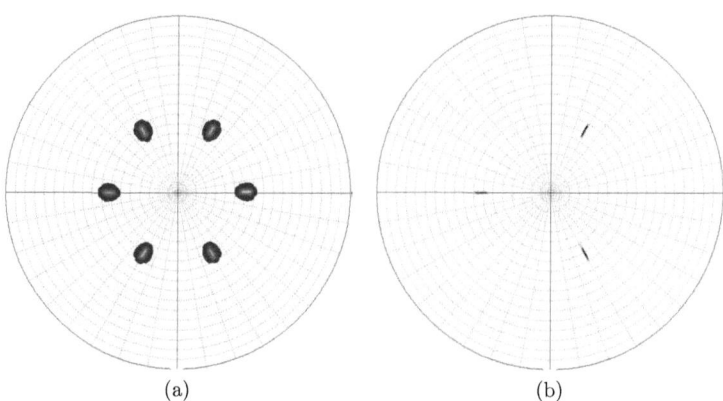

(a)   (b)

Figure 3.30: Pole-figures of Pb(Zr$_{0.4}$,Ti$_{0.6}$)O$_3$ grown on Pt/SrTiO$_3$ (111) with $2nm$ of TiO$_2$ seeding layer. (a) Pole-figure of the Pb(Zr$_{0.4}$,Ti$_{0.6}$)O$_3$ [110] directions measured at $2\theta = 31°.036$. The peaks are located at $35°$ which indicates PZT (111) growth. (b) pole-figure of the SrTiO$_3$ [110] direction at $2\theta = 32°.440$.

Figure 3.31: Dark field image of a $300nm$ thick $Pb(Zr_{0.4},Ti_{0.6})O_3$ (111) film on Pt (111) covered by $2nm$ TiO_2 prior to $Pb(Zr,Ti)O_3$ deposition.

3.10 Summary and Conclusion

3.10.1 Thin Films for the Subtractive Route

Epitaxial Pb(Zr$_{0.4}$,Ti$_{0.6}$)O$_3$ thin films have been deposited on single crystalline conductive Nb doped SrTiO$_3$ (100) substrates by sputtering. The treatment of the SrTiO$_3$ surface prior to the thin film deposition was found to be an important step to obtain Pb(Zr,Ti)O$_3$ films with the correct lattice parameters. The treatment leads to a complete TiO$_2$ termination of the SrTiO$_3$ (100) surface by a preferential removal of hydroxylated SrO terminations in buffered HF. Without the treatment, the lattice parameters of Ti rich Pb(Zr,Ti)O$_3$ compositions were much higher than compared with values found in the literature, and the in plane orientation was smeared out within 30°. The films were c-axis oriented with an estimated amount of a-domains of 14% in 400nm thick Pb(Zr$_{0.4}$,Ti$_{0.6}$)O$_3$ films and 25% in 600nm thick films. This relaxation mechanism in the surface normal direction with increasing film thickness was also observed in the literature. The Pb(Zr$_{0.4}$,Ti$_{0.6}$)O$_3$ films showed typical switching behavior of epitaxial c-axis oriented PZT where contributions from 90° domain walls are small. The d_{33} was measured as 45pm/V, the polarization charge was 65$\mu C/cm^2$, and the dielectric permittivity at zero field was 170. Using the LDG theory and available parameters from the literature, we were able to estimate the critical thickness, below which the polar state will be unstable at room temperature. This thickness was calculated as about 1nm.

3.10.2 Thin Films for the Additive Route

For the additive route, the epitaxial deposition of the sequence Pb(Zr,Ti)O$_3$/TiO$_2$/Pt was investigated on MgO (100) and SrTiO$_3$ (111) single crystalline substrates.

3.10.2.1 Deposition of Pt

The additive route is carried out on TiO$_2$ seeds on Pt (111) and on Pt (100). Epitaxial Pt (111) was obtained on single crystalline SrTiO$_3$ (111) substrates by a sputtering process leading to 50nm thick Pt film. Pole-figure measurements showed a twinning of Pt on SrTiO$_3$ (111) due to a fault in the A−B−C stacking sequence. There are two solutions of how Pt matches the (111) plane of SrTiO$_3$. One solution takes over the same crystallographic structure, and the other one is rotated by 60° with respect to the first one. TEM observation revealed a typical lateral twin size of 100nm. Pt (100) was obtained on single crystalline MgO (100) using a 85nm thick seeding layer of Ir on MgO. Direct deposition of Pt (100) on MgO (100) was difficult and not reproducible due to the low oxygen affinity of Pt, leading to mixed (111) and (100) orientations. Ir has a high oxygen affinity and grows epitaxially on MgO (100). Due to similar properties of Pt and Ir, epitaxial growth of Pt on Ir is achieved.

3.10.2.2 Deposition of TiO$_2$

For analysis, in order to gain enough X-ray diffraction intensity, 200nm thick TiO$_2$ films instead of only 2nm were deposited on the epitaxial Pt (111) and (100) layers. The depositions were carried out at 650°C (nominal). The TiO$_2$ crystallized in the tetragonal rutile phase. Even though the most stable plane in rutile TiO$_2$ is the (110) plane, (100) growth was obtained on both Pt orientations. Epitaxial mechanisms are responsible for this growth. Good lattice match is observed between the Ti atoms of the TiO_2 (100) plane with the (111) surface of Pt, where the c-axis of TiO_2 is parallel to the [110] axis of Pt. On Pt (100) the situation is more complex. Three types growth domains were observed. The abundance of the first type is higher than of the two others (higher intensity in pole figure measurement). The two other types are twins which can be brought into congruency when mirrored with respect to the a-axis of the first type's TiO_2 (100) plane. The twins are rotated by $\pm 28°.5$ with respect to the first type. To explain the epitaxial deposition, lattice matching considerations can only be taken into account for the first type. The c- and a-axis of TiO_2 are parallel to the Pt [110] directions. The c axis of TiO_2 matches reasonably well the Pt [110] direction, and the a axis can be brought into good match with Pt [110] after 3 unit cells of TiO_2.

3.10.2.3 Deposition of $Pb(Zr,Ti)O_3$

To obtain preferential growth of $Pb(Zr_{0.4},Ti_{0.6})O_3$ on TiO_2 seeds and not on the bare Pt, we have to be sure that $Pb(Zr,Ti)O_3$ is nucleation controlled also on Pt (100) (it is well known on Pt (111)). This was verified on a Pt (100) single crystal. Direct $Pb(Zr_{0.4},Ti_{0.6})O_3$ deposition, *i.e.* without application of a TiO_2 seed and/or $PbTiO_3$ starting layer, led to a strong island growth. Well (100) aligned square shaped features were observed, which indicates epitaxial growth. However, (111) triangles were observed as well. Hence, one can say, that $Pb(Zr_{0.4},Ti_{0.6})O_3$ is also nucleation controlled on Pt, but the perovskite phase nucleates better than on Pt (111). Depositing directly $PbTiO_3$ on Pt (100) surfaces led to $PbTiO_3$ (001) growth. Covering Pt (100) and (111) surfaces with $2nm$ of TiO_2 led to $Pb(Zr,Ti)O_3$ (111) growth in both cases. The orientation of TiO_2 dictates the orientation of $Pb(Zr,Ti)O_3$. If a $10nm$ thick $PbTiO_3$ starting layer was applied prior to $Pb(Zr,Ti)O_3$ deposition, the (111) orientation was still maintained and only a small fraction was (100) oriented.

Chapter 4

Fabrication of Ferroelectric Nano-Structures

This Chapter discusses the electron beam lithography process (EBL) for the subtractive and additive routes. The starting points are the substrates which were prepared as discussed in the previous Chapter, i.e. Pb(Zr$_{0.4}$,Ti$_{0.6}$)O$_3$ on Nb-doped SrTiO$_3$ (100) for the subtractive route, and for the additive route the substrates with the layer sequences 2nm TiO$_2$/50nm Pt/SrTiO$_3$ (111) and 2nm TiO$_2$/50nm Pt/85nm Ir/MgO (100). On these substrates, the EBL process is carried out.
First, an introduction to EBL is given, which presents the most important issues such as backscattering and proximity effects. The importance of the material below the resist layer is highlighted by means of Monte-Carlo simulations. The electron beam resists used in this work, polymethylmethacrylate (PMMM), and Calixarene are presented. The EBL process for subtractive route using epitaxial 50 to 200nm thick Pb(Zr$_{0.4}$,Ti$_{0.6}$)O$_3$ on SrTiO$_3$ (100) is then discussed. This process uses up to 200nm thick PMMA layers on Pb(Zr,Ti)O$_3$, Cr hard masking and lift-off. The size of single dots written into PMMA for different dot spacings and exposure times is discussed and the obtained Pb(Zr,Ti)O$_3$ features after dry etching are presented.
Then, the EBL is shown for the additive route. The EBL process using PMMA is discussed for the system 2nm TiO$_2$/50nm Pt/SrTiO$_3$ (111), and for the system 2nm TiO$_2$/50nm Pt/85nm Ir/MgO (100). Again, the size of individual dots as a function of dot spacing and exposure time is presented. The negative resist Calixarene was tested for the system 2nm TiO$_2$/50nm Pt/SrTiO$_3$ (111), and the obtained TiO$_2$ dot sizes are shown.
After the presentation of the electron beam lithography process, the obtained Pb(Zr,Ti)O$_3$ crystals after the deposition of 20nm Pb(Zr,Ti)O$_3$ with or without PbTiO$_3$ deposition prior to it are shown.
Special emphasis is given to the dry etching process.

4.1 Electron Beam Lithography (EBL)

The idea to use an electron beam for pattern transfer came soon after the discovery and construction of the first scanning electron beam microscope (SEM) by E. Ruska[283] in 1931. As the wavelength of the electrons decreases below 1Å, the limitation of the resolution due to the wavelength can be neglected. The interaction of the e-beam with organic compounds can be

78   CHAPTER 4. FABRICATION OF FERROELECTRIC NANO-STRUCTURES

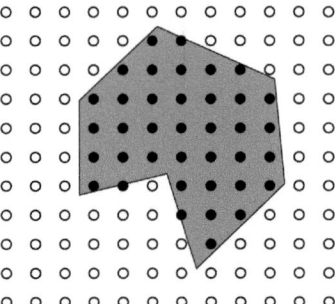

Figure 4.1: Raster of EBL writing process: Shapes are translated into arrays of dots (black) to be written. The proximity effect between neighboring dots leads to full exposure of the shape if a small enough step-size Δx (typically $5nm$) and the appropriate dwell-time t_{dw} is chosen.

of a bond scission (positive resist) or crosslinking (negative resist) nature. The formation of a contamination layer due to the interaction of the focused electron beam with the residual oil vapor is known since 1947.[284–286] So polymerized carbon patterns were successfully used as a hard mask to produce $8nm$ wide metal structures.[287] The positive PMMA resist was introduced by Hatzakis[288] in 1969 and is currently widely used. It soon became clear that the resolution is not limited by the primary beam diameter but by secondary electrons[287,289–291] generated by primaries through inelastic scattering and backscattering from the substrate. These secondaries show important interactions with matter due to their low energy ($< 50eV$[292]). Crosslinking or bonding scission takes place which in turn changes the dissolution ratio between exposed and unexposed volumes[293] in an appropriate solvent. Thus, calculating the tracks of secondary and backscattered electrons in a layered structure became important to predict and optimize the exposure. Monte Carlo simulations[289,294,295] can be used to calculate the interaction volume of electrons in the resist.

Due to the possibility to achieve high resolution and the direct writing ability, EBL became an important tool for the prototyping of the next generation of downsized devices such as MOS-FETs.[291,296–298] When the device feature size is reduced to less than $100nm$, quantum effects become noticeable,[299] in nano-structures such as single electron wells, quantum conduction effects in quantum wires,[300] Aharonov-Bohm rings,[301] in magneto-resistance of small rings and lines,[302] nano-island fabrication,[303] fabrication of metallic nanostructures[47,48] and well defined optical waveguides.[304] On a practical level EBL is used to fabricate masks for photo- and X-ray lithography.[305–308] The standard e-beam writer resembles very much that of a "normal" SEM, although TEMs have also been modified for EBL applications.[309] Additionally to an SEM, an EBL apparatus includes an electrostatic deflector (beam blanker), which allows the beam to be switched on and off during scanning.[310] The pattern to be written is translated into a raster of equidistant exposure-dots (see Figure 4.1, p.78). Two parameters govern the writing process: The step-size Δx (distance between the dots) and the dwell-time t_{dw} (exposure time) have to be chosen in a manner that the overlap (proximity effect) between the written dots leads to full exposure of the shape.

4.1. ELECTRON BEAM LITHOGRAPHY (EBL)

4.1.1 EBL Resists

The most important parameters for a resist in EBL are the sensitivity and the contrast. The sensitivity is the electron dose needed per area to develop fully the resist (see Table 4.1, p.79). A high sensitivity confers fast development and low process times. The contrast describes how fast a resist transforms from a partially to the fully developed state as a function of the electronic charge density. At too low charge densities, the resist is not fully developed. Increasing the dwell-time increases the charge density and gradually leads to a fully developed resist. The transition can be measured as the ratio of the exposed area which was fully developed to the total area of exposure. If the complete transition takes place in a narrow charge density range, the resist is called to have a high contrast. High contrast resists are desirable because they generally have more vertical side-walls. Figure 4.2, p.80 shows contrast curves for PMMA for different electron beam acceleration voltages. It can be seen that the contrast is slightly decreased at higher acceleration voltages. The onset of development takes place at the same dose (charge density), and the transition from 0 to 100% successfully developed area follows the same steepness for all acceleration voltages. The standard dose used for PMMA is $100 \mu m/cm^2$. From these graphs it can be seen that this dose leads to full development at all acceleration voltages.

Resist	Tone	Resolution (nm)	Sensitivity at $20kV$ [$\frac{\mu C}{cm^2}$]
PMMA	positive	10	100
EBR-9	positive	200	10
ZEP	positive	10	30
COP	negative	1000	0.3
SAL	negative	100	8
Calixarenes	negative	5	2500

Table 4.1: Examples of positive and negative EBL resists: Resolution limits and sensitivity (1997).[310]

4.1.1.1 Polymethylmethacrylate (PMMA)

Historically, PMMA was the first resist used for EBL[288] techniques and was also studied for deep UV-lithography.[311] Due to the high achievable resolution and acceptable sensitivity, PMMA is still used in EBL applications and techniques. PMMA is used as a positive resist. Bond scission takes place where the resist is exposed to the electron beam. The bond scission mechanism for PMMA is depicted in Figure 4.3, p.81. In the appropriate solvent (developer), exposed regions show much higher dissolution rates than unexposed areas. Hence, holes are obtained where the resist has been exposed to the electron beam. PMMA has also been observed to undergo cross-linking, but only at very high doses. This can also be seen in Figure 4.2, p.80 where the ratio of successfully exposed area decreases at very high doses ($1000\mu C/cm^2$). In this regime of high doses, PMMA starts to cross-link and is not dissolvable in the developer any more. In this case, the unexposed PMMA would dissolve faster than exposed regions. The resist works now in the negative regime. The achieved resolution in this regime is only $50nm$. Therefore, PMMA is almost exclusively used in the positive mode.[312]

Used as a positive resist, early experiments have been carried out at high electron energies (typically $50keV$) on thin and low density membranes to reduce backscattering from the underlying substrate.[287,313] Line widths of less than $20nm$ with a pitch of less than $60nm$ have

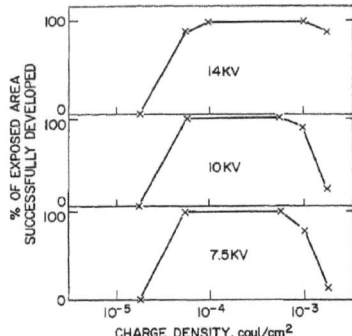

Figure 4.2: Contrast of the PMMA resist: Exposure characteristics of a $300nm$ thick PMMA film for different electron energies showing the amount of successfully developed matter as a function of the charge density. At the standard dose of $100\mu C/cm^2$, PMMA is fully developed at all acceleration voltages and works in the positive regime. For very high doses, PMMA becomes a negative resist: Cross-linking takes place leading to overexposed, unremovable resist areas.[288]

been achieved by these techniques.[314] The molecular weight of PMMA has an influence on the sensitivity. Low molecular weight PMMA is more sensitive than high molecular PMMA. The quality of lift-off processes can be enhanced by using a double layer PMMA resist where the layer below has a lower molecular weight than the top layer. During exposure, the successfully exposed area in the layer below is larger than in the top layer, and an undercut is achieved. Sometimes, lift-off processes suffer from the deposition of the hard mask on the side-wall of the patterned resist. These side-wall depositions are obstacles for a clean removal of the remaining PMMA. This is avoided with undercut resists. On thick silicon substrates, sub $40nm$ wide lines have been obtained[315] by this double layer technique.

4.1.1.2 Calixarenes

Calixarenes[317, 318] are cyclic condensation products made up of para-substituted phenols and formaldehyde. They have a chalice-like structure. A typical structure used in EBL, and in the scope of this dissertation is a Calixarene molecule with a diameter of $1nm$[319] shown in Figure 4.4, p.82. It was discovered as an ultrahigh resolution negative resist for EBL. Due to its phenolic nature, it shows good halide dry etching resistance.[320, 321] The etching rate in CF_4 was found as $10nm/min$, which is about the same as for Si but four times smaller than PMMA. In general, Calixarenes are low sensitivity resists with sensitivities[322] ranging from $700\mu C/cm^2$ to $7mC/cm^2$ (compared to $100\mu C/cm^2$ for PMMA) showing sharp edges in the sub-$100nm$ region.[323] The sensitivity seems to increase with increasing number of phenolic units.[318] The mechanism responsible for the crosslinking has not been resolved as yet.[318] $10nm$ wide lines have been obtained on Ge using $30kV$ acceleration voltage. On Si substrate it was possible to produce dots with $15nm$ diameter having a $35nm$ pitch between them.[321, 322, 324] Experiments have been carried out at high energy ($50kV$) and it was found that backscattering is be the limiting factor

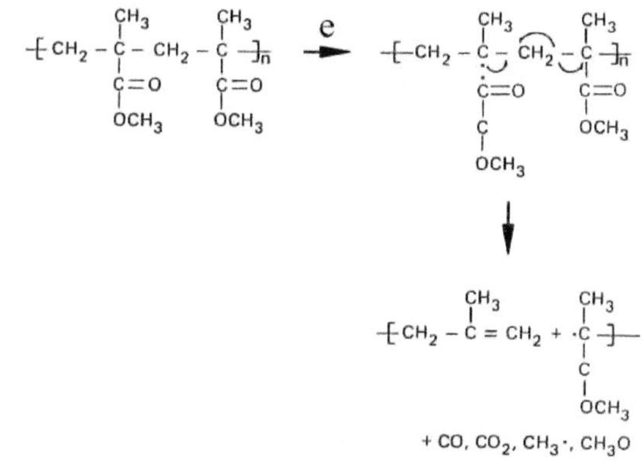

4.1. ELECTRON BEAM LITHOGRAPHY (EBL)

$$\begin{array}{c} CH_3 \quad\quad CH_3 \\ | \quad\quad\quad | \\ -[CH_2-C-CH_2-C-]_n \\ | \quad\quad\quad | \\ C=O \quad C=O \\ | \quad\quad\quad | \\ OCH_3 \quad OCH_3 \end{array} \xrightarrow{e^-} \begin{array}{c} CH_3 \quad\quad CH_3 \\ | \quad\quad\quad | \\ -[CH_2-C-CH_2-C-]_n \\ | \quad\quad\quad | \\ C=O \quad\quad C=O \\ | \quad\quad\quad | \\ C \quad\quad\quad OCH_3 \\ | \\ OCH_3 \end{array}$$

$$\downarrow$$

$$-[CH_2-\underset{\underset{OCH_3}{|}}{\underset{|}{C}}=CH_2 + \cdot\underset{\underset{OCH_3}{|}}{\underset{|}{C}}-]-$$

$$+ CO, CO_2, CH_3\cdot, CH_3O$$

Figure 4.3: The backbone of PMMA chains is broken by interaction of low-energy secondary electrons with the covalent binding of the side groups.

for dot separation. Due to the low sensitivity of Calixarene resists, combined optical (high throughput)/EBL techniques have been investigated and a MOSFET with a gate length of $20nm$ was achieved. Replacing the methyl-group by chloromethyl increased the sensitivity by a factor of 10 due to easy decomposition of the Cl-C bond.[322,325] Decreasing the electron beam energy increases the sensitivity according to the Bethe equation for energy absorbency, and the minimum dot size on Si was found to be $10nm$ regardless the beam energy.[319] Higher resolution was obtained using the lower sensitive resist due to reduced influence of backscattering. Low-energy electron beam techniques have two advantages: increases in sensitivity (low process time) and backscattering reduction due to the small penetration depth of electrons. But even in this process, the lower resolution limit was only $10nm$ lines.[326] The solubility ratio between exposed and unexposed regions was found to be very high. Yasin et al.[327] have shown that ultrasonically-assisted development further increases the resolution. Full development was observed after $30s$, and a development time of $60min$ did not remove the exposed features.[316] The resist could only be stripped using an oxygen plasma. This is an attractive characteristic, which further underlines the high resistance to dry etching, and opens the possibility to pattern transfer down to $10nm$ into SiO_2 for gate processing.[297,312,325,328] As shown before, the resolution limit on Si substrates was found to be $10nm$. Si is a low density material providing small backscattering. EBL with Calixarene was also used to produce magnetic features from Fe films. The resolution on Fe was found to be $50nm$[329] at $50kV$.

4.1.2 Resolution Limits

Although electron beam diameters of a few nanometers can be achieved,[330,331] it is unfortunately not the limiting factor for resolution in EBL. In fact, the incident high energy primary electrons (PE) undergo several scattering events on their way through the resist into the substrate, which arise from the interaction with atomic shells. Elastic low loss, or low angle scattering (called forward scattering in EBL), leads to a slight broadening of the incident beam. More impor-

82   CHAPTER 4. FABRICATION OF FERROELECTRIC NANO-STRUCTURES

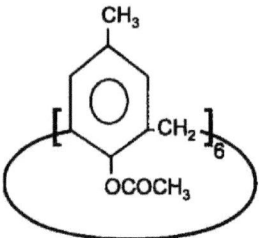

Figure 4.4: The Calixarene molecule[316] used in this work. The repetitive unit is repeated 6 times to form a ring. The notation is 4-Methyl-1-Acetoxycalix[6]aren.

tant in EBL are inelastic high energy loss, wide angle scattering events, where electrons are deviated far away from the incident beam. High angle scattering is an important problematic when performing EBL on high density substrates,[332] for example when working on metallic thin films. Backscattered electrons from the metal film reach the overlaying resist film far away from incident beam. Figure 4.5, p.83 shows a Monte-Carlo simulation[333] of an e-beam hitting into bulk PMMA (a) and into a PMMA layer on Pb(Zr,Ti)O_3/SrTiO_3 (b). In the low density PMMA material the electrons travel far into the bulk (several tens of micrometers). With a higher density Pb(Zr,Ti)O_3 film on an underlying SrTiO_3 substrate, the scattering events are concentrated near the surface and large backscattering takes place into the PMMA. To reduce the negative impact of backscattering, either low energy EBL[326] (less than $5kV$) or high energy EBL[309] (greater than $40kV$) can be used. In the former case the energy is chosen in a way that the e-beam just reaches the bottom of the resist layer, but not the substrate. In the latter case, by using a high energy beam, the volume of back scattered electrons is displaced away from the resist into the substrate.

On their travel through the resist, the PE generate secondary electrons (SE, electrons with energies less than 50eV). The SE yield depends on many factors,[332] and it is generally higher for high atomic number substrates. The most probable energy of SE ranges from 1 to $5eV$ and their average mean free path in polymers is around $10-20nm$.[292,334] In general, the probability of electron-molecule interaction is suddenly decreased as the electron energy increases.[321] These low energy SE interact with the covalent bindings of organic matter (binding scission and crosslinking) as they have about the same energy. The secondaries and not the primaries are responsible for the exposure of the resist. The exposure by backscattered electrons far away from the point of incidence is responsible for what is called the proximity effect. In the case of an isolated pattern, the dose is too small to completely develop the resist there. However, in the case of dense patterns, the dose from neighboring features may add to fully develop the area between patterns. In very dense structures this effect can be corrected[291,335–337] by reducing the dose at the border of neighboring patterns, for example by decreasing the dwell-time or by increasing the step-size. The proximity effect is described by an exposure intensity distribution curve which can be measured experimentally. Significant exposure of PMMA takes place even $500nm$ away from incidence for a $600nm$ thick PMMA film on Si.[335,336]

Monte-Carlo simulations have been performed for every layer sequence used in this work (see Figure 4.6, p.87). The calculations have been carried out for PMMA/Pb(Zr,Ti)O_3/SrTiO_3 (a),

4.1. ELECTRON BEAM LITHOGRAPHY (EBL)

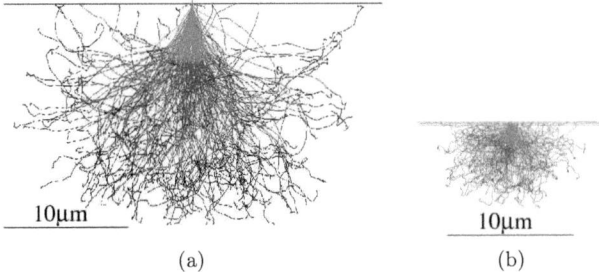

Figure 4.5: Monte-Carlo simulations of a $20kV$ electron beam hitting into bulk PMMA (a) and a $150nm$ thin PMMA film on $200nm$ Pb(Zr,Ti)O$_3$ on SrTiO$_3$.

PMMA/Pt/SrTiO$_3$ (b), Calixarene/Pt/SrTiO$_3$ (c), and PMMA/Pt/Ir/MgO (d). The simulations have been done for acceleration voltages of $10kV$ (left column) and $20kV$ (right column). A backscattering coefficient (BS) is indicated in the figures. It can be seen that BS is lowest for PMMA on Pb(Zr,Ti)O$_3$/SrTiO$_3$. This is due to the low densities of the underlying layers. Ir and Pt have high densities of $22g/cm^3$ compared to Pb(Zr,Ti)O$_3$ ($8.1g/cm^3$), SrTiO$_3$ ($5.1g/cm^3$) and PMMA itself ($2.6g/cm^3$). For $200nm$ PMMA, a $10kV$ beam hardly develops the whole thickness. For simulations on metallic Pt layers (b) − (d), even though the overall interaction volume is smaller for a $10kV$ beam than for a $20kV$ beam, backscattering at $10kV$ is concentrated around the vicinity of the incident position leading to full development there. This concentration leads to a larger spot size at $10kV$ compared to a $20kV$ beam. At an acceleration voltage of $20kV$, backscattering is spread over a large volume. This decreases the local dose and therefore, not all of the interaction volume is fully developed. Hence, the $20kV$ beam is preferable. On $50nm$ Pt, the higher energy beam shows even less backscattering due to the displacement of the beam further away from the surface. When the thickness of the high density material is increased, as it can be seen in the case of Pt/Ir (d), the increase of the electron beam acceleration does not help any more to decrease the effect of backscattering.

Besides beam energy, there are other factors for the choice of the lithographic process. Since the pattern transfer in e-beam lithography is a slow process, one prefers to write the smallest possible area. The features in our case are dots. It takes much less time to write the dots than to write the area around the dots. After developing a positive resist like PMMA, we obtain holes where the resist has been exposed to the electron beam. In order to keep the process short, the mask in the positive resist has to be inverted. This is done by Cr hard masking and lift-off. The insertion of this mask-inversion step is time consuming, and might have consequences to the underlying surface after the final removal of the hard mask. However, the Cr mask is hard, *i.e.* resistant to the etching process. It can therefore be used to etch the underlying layer.

A simpler and faster process is the use of negative resists which directly produce a mask. In this case, the mask is organic, with generally poor resistance to the etching process. So, we can formulate the requirements as:

- Is mask inversion needed ?

- Do we need a harder mask as offered by the patterned resist ?

Subtractive route	Additive route
Pb(Zr$_{0.4}$,Ti$_{0.6}$)O$_3$ [3]/SrTiO$_3$ (100)	TiO$_2$ [1]/Pt [6]/SrTiO$_3$ (111)
	TiO$_2$ [1]/Pt [6]/Ir [5]/ MgO (100)

Table 4.2: Samples with epitaxial films as prepared to perform EBL. The number in brackets designates the sputtering deposition condition in Table 3.1, p.43.

In our case, we encounter two different scenarios. For the subtractive route, we used a mask that resists better to the etching process than PMMA, and the Pb(Zr,Ti)O$_3$ film is much thicker than the PMMA layer. For the additive route only a $2nm$ thick TiO$_2$ layer has to be patterned. Therefore a negative electron beam resist is sufficient.
Before describing the two routes, the experimental methods for the fabrication are described.

4.2 Experimental Methods for the Lithography Process

4.2.1 Preparation of Resist Thin Films for EBL

We use two different resists:

- Positive resist: PMMA (Poly-Methyl-Methacrylate) (see Figure 4.3, p.81)
- Negative resist: Calixarene (4-Methyl-1-acetoxycalix[6]aren) (see Figure 4.4, p.82)

The $4\% - vol.$ PMMA-anisol solution was bought from YMC-Yield Management, Rapperswil, Switzerland. The solution was further diluted with Anisole to get a $1\% - vol.$ PMMA solution. The Calixarene solution was bought from Allresist GmbH, Germany.
The solutions were spun onto the films and rotated for $45s$ at $6000rpm$ (PMMA) or $4000rpm$ (Calixarene). The bake was performed by placing the sample on a preheated hot-plate ($175°C$) during $20min$. The thickness of one PMMA deposition was exactly measured by means of a Nanospec system that uses non-contact, spectro-reflectometry (measurement of the intensity of reflective light as a function of incident wavelength) to determine the thickness of transparent films on substrates. The thickness was measured as $54nm$. It was possible to deposit two consecutive layers of PMMA (after baking out the first one), and the thickness was exactly doubled. Systematic AFM measurements showed that the thickness was highly reproducible. AFM measurements on Calixarene revealed a thickness of 50nm. It was not possible to deposit consecutive layers of Calixarene because the former layer is completely dissolved during the second spin coating process.

4.2.2 Evaporation

Thermal evaporation is used to fabricate the Cr hard masks for the lift-off process of PMMA. This process is strongly directional, *i.e.* depositions on the side-wall of PMMA cavities are avoided. Side-wall deposition of the hard mask may limit the access of the solvent to the PMMA during the lift off process. The Cr source consists of a W rod covered by $0.5mm$ of Cr. Resistive heating of the W rod leads to the evaporation of Cr which is deposited on the substrate. The deposition takes place at a distance of $30cm$ and the deposition rate is controlled at $0.5nm/s$. The deposition is started when the base pressure is below $5 \times 10^{-7} Torr$.

4.2. EXPERIMENTAL METHODS FOR THE LITHOGRAPHY PROCESS

	mp [$°C$]	bp [$°C$]
$PbCl_4$	-15	50
$TiCl_4$	-24	136
TiF_4	284	subl.

Table 4.3: Melting and boiling point[274] at room temperature and atmospheric pressure of probably "unstable" compounds. TiF_4 sublimates at atmospheric conditions.

4.2.3 Dry Etching in an ECR/RF Reactor

As the dimensions of a ferroelectric capacitor shrink, the state of the surface becomes more and more important. This is due to the increased surface/volume ratio (see Chapter 2, p.9). Dry etching may cause surface defects which lead to domain pinning and so to reduced piezoelectric activity.

It is generally reported that dry etching processes deteriorate the properties of $Pb(Zr,Ti)O_3$ cells. During inductively coupled plasma (ICP) etching using $Ar-Cl_2-C_2F_6$ gas mixtures, a damaged surface layer of $10nm$ was created on the side-walls[338] of the PZT features. This led to an increase of the coercitive field and a decrease in the switching polarization. After cleaning the side-wall in a wet etching process, the properties were restored, and leakage currents decreased.[338] Etch-processes which provide high etch rates and good etch anisotropy for $Pb(Zr,Ti)O_3$ and SBT, use the highly reactive Cl^- and F^- species present in hydro-chloro-fluoro-carbons or CCl_4/CF_4 gas mixtures. Surface adsorbed chemicals and ion bombardment-induced defects in the near surface region change the interface character, and therefore the properties of the film.[339] As the device feature size decreases, the physical damage and chemical residue effects become much more severe. The solid residues on the etched surface are mainly fluorides.[340] A shift in the ferroelectric loop has been observed comparing loops before and after dry etching.[339] High pressures are not desirable for $Pb(Zr,Ti)O_3$ etching because of increased re-deposition.[341]

In this work, etching took place in a dual frequency ECR/RF reactor (electron cyclotron resonance ion gun combined with RF biasing of the substrate).[340,342] The setup is similar to the one used by Boukari et al.[343] The ECR ion gun is operated at $2.45GHz$ and the frequency of the bias-signal to the substrate is $13.56MHz$. The used etchants are Ar, CF_4 and CCl_4. This gas mixture assures a good balance between pure chemical and pure physical Ar etching. This system is able to produce a plasma at extremely low pressures. As such, etching can be done at pressures lower than $0.08mTorr$. At such low pressures, no re-deposition, fencing or contamination takes place in contrast to standard Ar milling[344] which runs at pressures at around $10mTorr$. The energy of the ions is not only controlled by the RF bias to the substrate but also by two grid voltages for extraction (V_{ext}) and acceleration (V_{acc}). Table 4.4, p.86 shows the etching rates obtained on several materials.

Chemical etching considerably increases the etching rate if the atoms in the film form volatile chloride or fluoride compounds with the etchants. In order to estimate the volatility of the compounds, the melting and boiling points are taken into account. The compounds of Pt ($PtCl_2$, $PtCl_4$, PtF_4), Pb ($PbCl_2$, PbF_2, PbF_4), Ti ($TiCl_2$, $TiCl_3$, TiF_3) and Zr ($ZrCl_2$, $ZrCl_4$, ZrF_4) have melting and boiling points above $300°C$ at atmospheric pressure and can be assumed to be stable at low pressures. Only $PbCl_4$, $TiCl_4$, TiF_4 show much lower values[274] (see Table 4.3, p.85). The formation of these compounds during etching is responsible for the high etching rate when adding CF_4 and/or CCl_4 (see Table 4.4[14-17], p.86, for $Pb(Zr,Ti)O_3$ and [18 − 22] for TiO_2).

#	Material	Ar [sccm]	CF$_4$ [sccm]	CCl$_4$ [sccm]	Bias [W]	V_{ext} [V]	V_{acc} [V]	Rate [nm/min]
[1]	Cr	10	40	0	40	0	0	1.8
[2]	Cr	10	40	0	20	0	0	0.7
[3]	Cr	20	0	20	40	0	0	3
[4]	Cr	20	0	0	20	300	300	2.3
[5]	Cr	20	20	20	40	0	0	3.3
[6]	Cr	8	8	8	40	150	150	2
[7]	Pt	0	0	18	20	0	0	0
[8]	Pt	10	40	0	20	0	0	1.8
[9]	Pt	10	40	0	40	0	0	5
[10]	Pt	20	0	0	40	0	0	2
[11]	Pt	20	20	0	40	0	0	4.3
[12]	Pt	20	0	20	40	0	0	3.3
[13]	Pt	20	0	0	20	300	300	17
[14]	Pb(Zr,Ti)O$_3$	20	20	20	40	0	0	10
[15]	Pb(Zr,Ti)O$_3$	20	20	20	40	60	60	12
[16]	Pb(Zr,Ti)O$_3$	8	8	8	40	150	150	14
[17]	Pb(Zr,Ti)O$_3$	20	0	0	20	300	300	35
[18]	TiO$_2$	10	40	0	40	0	0	7
[19]	TiO$_2$	20	0	0	10	0	0	0.2
[20]	TiO$_2$	20	20	0	40	0	0	3.1
[21]	TiO$_2$	20	0	20	40	0	0	4.8
[22]	TiO$_2$	20	20	20	40	0	0	6.5
[23]	TiO$_2$	20	0	0	20	300	300	2
[24]	PMMA	20	0	0	40	0	0	8
[25]	PMMA	20	0	20	40	0	0	32
[26]	PMMA	20	20	0	40	0	0	32
[27]	Calixarene	10	40	0	0	0	0	25
[28]	Calixarene	20	20	20	0	0	0	35

Table 4.4: Etching conditions.

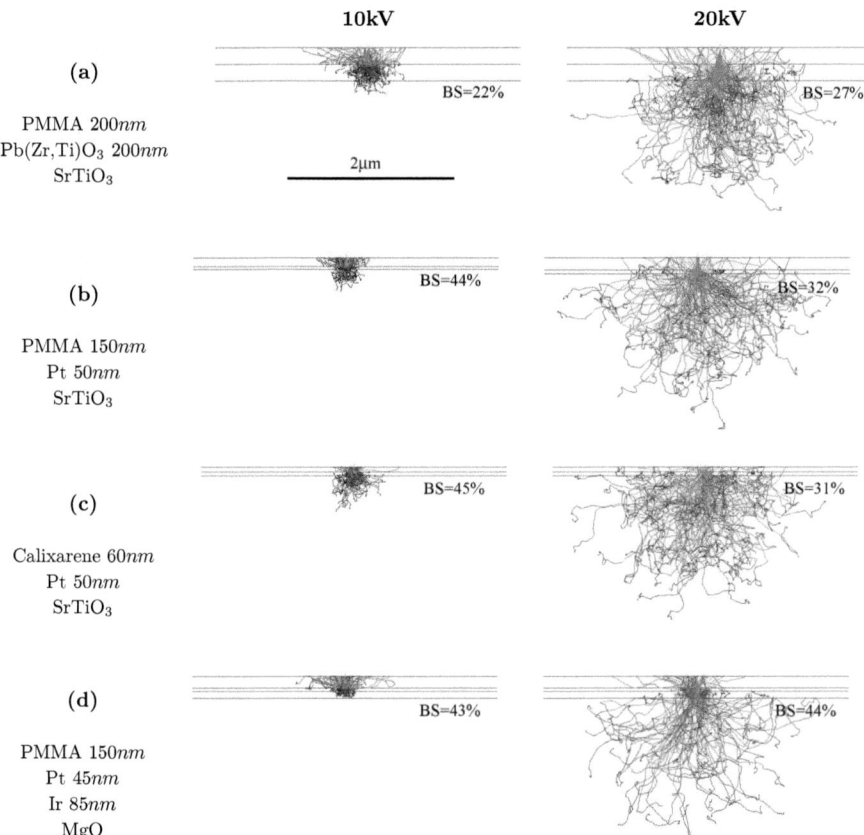

Figure 4.6: Monte-Carlo simulations of an e-beam hitting into different substrates at $10kV$ (left) and $20kV$ (right). The backscattering coefficient is indicated as "BS".

CHAPTER 4. FABRICATION OF FERROELECTRIC NANO-STRUCTURES

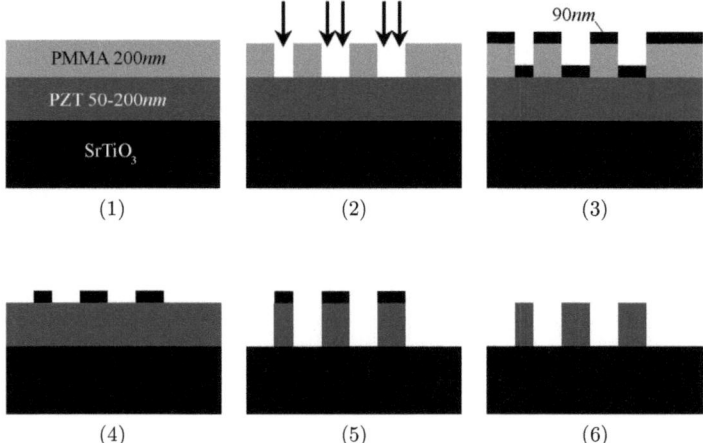

Figure 4.7: Process flow for the subtractive method. (1) starting point: SrTiO$_3$ (100) substrate with epitaxial Pb(Zr$_{0.4}$,Ti$_{0.6}$)O$_3$ (001), covered with the positive electron beam resist PMMA, (2) exposure and development of PMMA, (3) Cr evaporation, (4) lift-off in acetone, (5) dry etching in ECR reactor, (6) removal of remaining Cr in an aqueous solution of ceric ammonium nitrate and perchloric acid.

4.3 The Subtractive Route[1]

In the subtractive route we used SrTiO$_3$ (100) single crystalline substrates on which $50nm$ to $200nm$ thick Pb(Zr$_{0.4}$,Ti$_{0.6}$)O$_3$ films were deposited (see Table 3.1[3], p.43 for deposition conditions of Pb(Zr,Ti)O$_3$). In this route, the Pb(Zr,Ti)O$_3$ film is patterned. The process flow for this route is given in Figure 4.7, p.88. (1) PMMA of the desired thickness is spun on, (2) electron beam exposure and development leads to PMMA with holes that define the features, (3) Cr is evaporated, and the lift-off performed in acetone (4), (5) the sample is submitted to a dry etching process, and finally, (6) the Cr is removed.

A $2\% - vol.$ PMMA solution in anisole was used for spin-coating of PMMA on Pb(Zr,Ti)O$_3$. The thickness achieved at $6000rpm$ during $45s$ was $54nm$. The bake out was done on a hot plate at $175°C$ during $30min$. Multiple coating was possible, with the thicknesses just adding. The thickness of the PMMA should exceed at least twice the required thickness of the Cr hard mask. Under the used etching conditions ([5] for Cr and [14] for PZT in Table 4.4, p.86), Cr vanishes three times slower than PZT. Hence, for a $200nm$ thick Pb(Zr$_{0.4}$,Ti$_{0.6}$)O$_3$ film, at least $66nm$ Cr are required. We used $90nm$ Cr which requires a $200nm$ thick PMMA layer.

In a first attempt, a pattern was written into $200nm$ PMMA from a predefined mask. The designed features were triangles, circles, squares and rectangular shapes. In this fabrication method, the shapes have been defined by writing a grid of dense dots (see Figure 4.1, p.78). The chosen step-size was $5nm$, and the beam current was adjusted in order to reach a dose of $100\mu C/cm^2$, at which full development of PMMA takes place. The distance between features was as small as $200nm$. Figure 4.8(a), p.89 shows non-contact AFM scans on the PMMA after e-beam exposure and development. The triangular and square shaped features were well

4.3. THE SUBTRACTIVE ROUTE[1]

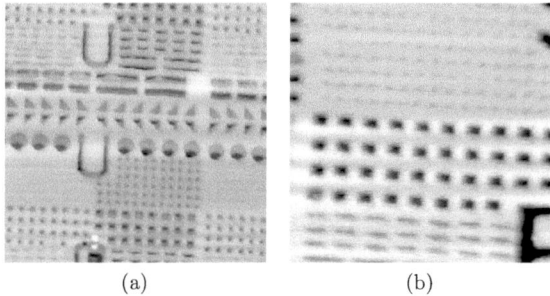

Figure 4.8: AFM scan on PMMA after electron-beam exposure and development on $200nm$ PMMA/$200nm$ Pb(Zr,Ti)O$_3$/SrTiO$_3$ (100). (a) $5 \times 5\mu m$, (b) $2 \times 2\mu m$.

defined at that stage. Initially square shape designed features were still square shaped after development. From the AFM scan it is not clear whether the smallest $50nm$ designed dots have been fully developed down to the bottom. Some distortion in the PMMA material seemed to have happened around long rectangular features with small separation. The resolution suffered considerably after translating the negative features into positive ones by evaporating $90nm$ Cr and lift-off in acetone. Figure 4.9, p.89 shows SEM images of the Cr hard mask obtained after evaporation through the PMMA mask, and after lift-off (removal of PMMA in acetone). The triangular shaped features which were well defined holes in PMMA, now show blurred edges and the rectangular long features show a non uniform filling (b) leading to blurred side walls. Features with one lateral size of $50nm$ (Figure 4.8(b), p.89 above and below center) could not be translated into a Cr hard mask. They are missing (white circle in Figure 4.9(a), p.89). This could be due to incomplete access of Cr through the hole or due to an insufficient dose during the exposure of small features leaving PMMA at the bottom of the hole.

For dry etching of Pb(Zr,Ti)O$_3$, different conditions were used (see Table 4.4[14–17], p.86). The best condition was the condition at equal CF$_4$, CCl$_4$ and Ar content without using an additional acceleration and extraction voltage [14]. In that case, a good balance between chem-

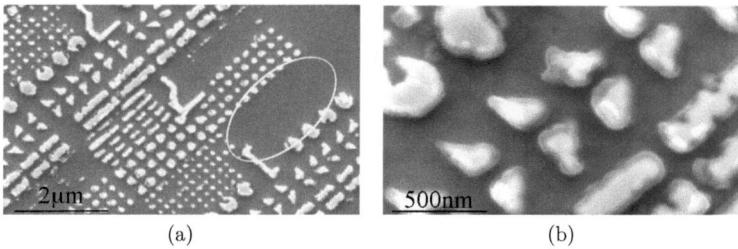

Figure 4.9: SEM images on $100nm$ Cr/Pb(Zr,Ti)O$_3$/SrTiO$_3$ (100) after evaporation and lift-off in acetone. (a) large scale view of the features. The white circle indicates missing features with a lateral size of designed $50nm$. (b) triangular and rectangular shaped features.

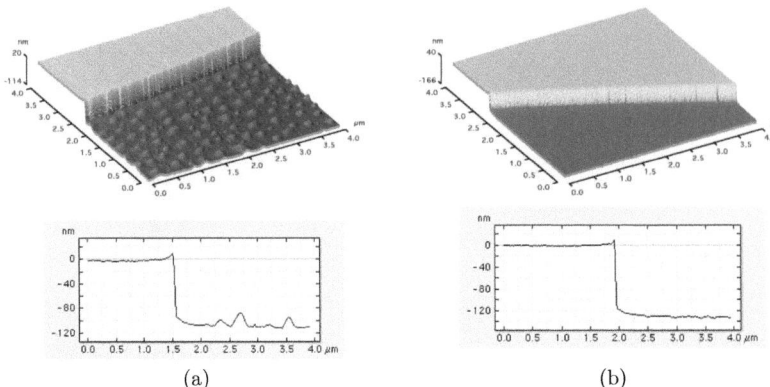

Figure 4.10: Dry etching of PZT. (a) using only Ar and $300V$ extraction and acceleration voltage (Table 4.4[4], p.86), (b) using CF_4, CCL_4 and Ar (Table 4.4[14], p.86). The etching rate in the pure Ar process was $35nm/min$ against $10nm/min$ in the other case. Both methods show sharp side-wall definition as evidenced by the AFM line scans.

ical and physical etching was obtained leading to uniform etching and smooth surfaces. Figure 4.10(b), p.90 shows an AFM scan on a partially etched $Pb(Zr,Ti)O_3$ surface, where no acceleration and extraction voltages have been applied during the etching process. A flat surface is obtained. In contrast, (a) shows the partial etching of $Pb(Zr,Ti)O_3$ using only Ar and $300V$ acceleration and extraction voltage. In (a) the etching rate was $35nm/min$ and in (b) it was $10nm/min$. In (a), the surface shows non uniform etching with $20nm$ high spikes appearing. This was observed for all etching conditions using an acceleration and extraction voltage. Hence, condition [14] was chosen for dry etching of PZT.

Figure 4.11, p.90 shows the features obtained after dry etching. The side-walls of the features are rounded. This may be due to the inhomogeneous thickness of the previously deposited Cr mask

Figure 4.11: SEM image on patterned, initially $200nm$ thick $Pb(Zr,Ti)O_3$ on $SrTiO_3$ (100) after dry etching and removal of Cr.

4.3. THE SUBTRACTIVE ROUTE[1] 91

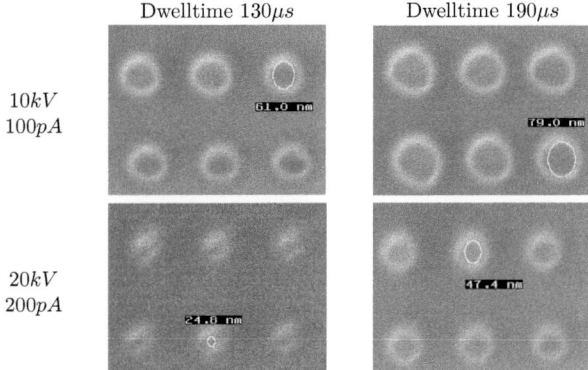

Figure 4.12: SEM images of dots created into $200nm$ thick PMMA at different electron beam acceleration voltages and dwell-times. Acceleration voltage: $10kV$ with $106pA$ (up) and $20kV$ with $180pA$ (down), dwell-time: $130\mu s$ (left) and $190\mu s$ (right). The dots have a separation of $200nm$ from center to center.

features. Indeed, AFM measurements have shown that the Cr features were thinner towards the edge of a pattern. During etching, the Cr can be completely removed at the border of a feature leaving unmasked $Pb(Zr,Ti)O_3$. Sometimes, the separation between the features is incomplete. $50nm$ patterned features are missing and the smallest lateral feature size obtained was $100nm$.
As a conclusion, this method led to features with the smallest lateral size being $100nm$. Features smaller than $150nm$ showed blurred out borders. The transfer of Cr was insufficient for features with a lateral size smaller than $50nm$.
In order to access the smallest possible feature size, single dots were written at variable dwelltimes and constant spacing. Figure 4.12, p.91 shows holes obtained in PMMA at acceleration voltages of $10kV$ ($100pA$) (top) and $20kV$ ($200pA$) (bottom) and for two different dwell-times: $130\mu s$ (left) and $190\mu s$ (right). The distance from dot-center to dot-center was $200nm$. For such close distances, the proximity effect and overlap of backscattering becomes an important issue.

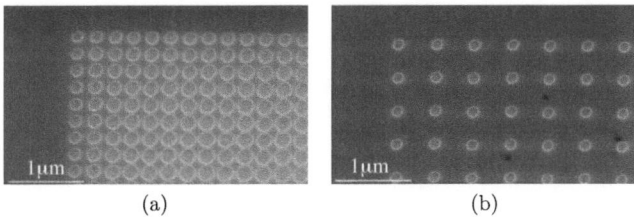

Figure 4.13: SEM images of holes obtained on $100nm$ PMMA/$100nm$ $Pb(Zr,Ti)O_3$/SrTiO$_3$ (111). The dots were written at $20kV$ acceleration voltage, $200pA$ and $300\mu s$ dwell-time. The distance from center to center was $250nm$ in (a) and $500nm$ in (b).

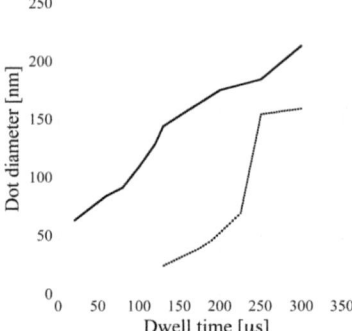

Figure 4.14: Graph showing the dot diameter obtained in $200nm$ PMMA/$200nm$/Pb(Zr,Ti)O$_3$/SrTiO$_3$ (100) for two different dot spacings as a function of dwell-time. The exposition condition was $20kV$ acceleration voltage and $200pA$. Plain line: dots with a spacing center-center of $250nm$, dotted line: spacing of $500nm$.

It can be seen that the lower acceleration voltage led to larger dots. This is explained by the interaction volume which is displaced away from the surface for higher acceleration voltages, leading to a more spread out backscattering and hence to a lower dose from backscattering (see also Figure 4.6(a), p.87). As expected, a larger dot-size is obtained at longer dwell-times. Generally, the measured dot size is much larger than calculated from the dose (which is proportional to the dwell-time). This is due to the additional proximity effect and backscattering. The influence of the proximity effect is highlighted in Figure 4.13, p.91. This figure shows the border region of a field covered with single dots obtained using the same conditions, $20kV$ and $200pA$. The difference is the spacing between the dots, which is $250nm$ in (a) and $500nm$ in (b). In both cases, the dots were written at a dwell-time of $300\mu s$. In the case of the larger spacing (b) all dots show a uniform size. This means that at this spacing, the proximity effect is negligible.

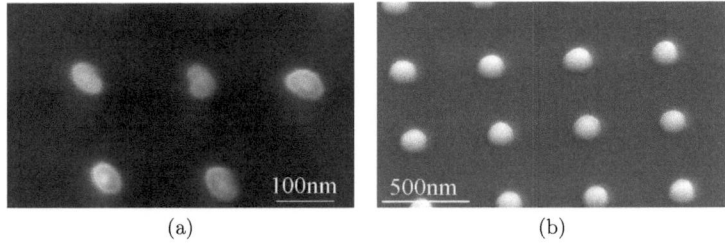

(a) (b)

Figure 4.15: SEM images of the smallest dots obtained by the subtractive method using a $100nm$ thick Pb(Zr,Ti)O$_3$ film on SrTiO$_3$. (a) $50nm$ features after Cr evaporation and lift-off, (b) $80nm$ features after dry etching.

4.4. THE ADDITIVE ROUTE[2]

At a dot distance of 250nm (a) the proximity effect can clearly be seen. The dots are generally larger than in (b) but dots at the border are significantly smaller than dots towards the inner part of the written region. This is because dots at the border have less neighbors from which they receive an additional dose (proximity effect). The influence of the dot spacing (proximity effect) and dwell-time on the dot diameter is depicted in Figure 4.14, p.92. This figure shows the obtained dot diameter in 200nm PMMA/200nm Pb(Zr,Ti)O$_3$/SrTiO$_3$ (100) after development. The exposition parameters were 20kV acceleration voltage and 200pA beam current. For dots spaced by 250nm center-center, the smallest dot had a diameter of 60nm at a dwell-time of 20μs. Below this dwell-time, the dose was insufficient and no dots have been obtained after development. Above this threshold, the dot diameter increased linearly with the dwell-time. For this dot spacing, at a dwell-time of 300μs, PMMA started to be fully developed even between the dots. This proximity effect is greatly reduced when increasing the dot distance from 250nm to 500nm (dotted line). Here, at 200μs dwell-time, the dot-size was 50nm compared to 175nm for dots separated by 250nm. The smallest dot obtained in PMMA was 25nm. For 500nm spaced dots, a maximum dot diameter of 160nm exists.

As a conclusion, smaller feature sizes have been achieved using the single dot method. The smallest Cr feature obtained on Pb(Zr,Ti)O$_3$/SrTiO$_3$ was 50nm (Figure 4.15(a), p.92). However, after dry etching these small dots disappeared due to insufficient Cr deposition into small cavities. The smallest features obtained after dry etching had a diameter of 80nm (b).

4.4 The Additive Route[2]

In the additive route, the deposition of PZT is done on a Pt layer covered with TiO$_2$ seed islands. For the fabrication of the seeds, a 2nm thick TiO$_2$ seeding layer is etched instead of Pb(Zr,Ti)O$_3$. A subsequent Pb(Zr,Ti)O$_3$ deposition then leads to preferential nucleation of Pb(Zr,Ti)O$_3$ crystals on the TiO$_2$ seed islands and not on Pt. We tested the method on Pt/SrTiO$_3$ (111) and on Pt/Ir/MgO (100). As only 2nm of TiO$_2$ have to be etched, negative organic resists can be used as mask instead of the positive PMMA resist, which needs Cr hard masking and lift-off. Figure 4.16, p.94 shows the process flow using the negative Calixaren resist. There is no lift-off process. (1) The Calixarene resist is spun on and baked at 175°C for 30min. This yielded a 60nm thick Calixarene film. Multiple depositions were not possible because subsequent Calixarene depositions dissolved the previous layer. (2) Exposure and development leads to dissolution of non-exposed material leaving back the organic mask. (3) Dry etching of TiO$_2$ was done at the condition shown in Table 4.4[18], p.86. TiO$_2$ is etched at a rate of 7nm/min compared to 25nm/min for the Calixarene resist ([27]). The remaining Calixarene was then removed in an oxygen plasma (4). Finally, Pb(Zr$_{0.4}$,Ti$_{0.6}$)O$_3$ is deposited with a nominal thickness of 20nm with or without a PbTiO$_3$ seed layer prior to Pb(Zr,Ti)O$_3$ deposition.

4.4.1 Fabrication of TiO$_2$ Dots using PMMA

To illustrate the process using the positive resist PMMA and Cr hard masking, step (2) can be replaced with the steps (2–4) from the process flow of the subtractive route (Figure 4.7, p.88). The process with PMMA was tested on 50nm Pt/SrTiO$_3$ (100) and on 50nm Pt/85nm Ir/MgO (100). The important difference in terms of electron beam lithography is the larger thickness of high density material (Pt plus Ir) below PMMA, which leads to large backscattering effects in the latter case. Cr hard masking and lift-off for the dry etching of 2nm TiO$_2$ would theoretically require a Cr layer of only several nm. However, it was found that such thin Cr layers were not

CHAPTER 4. FABRICATION OF FERROELECTRIC NANO-STRUCTURES

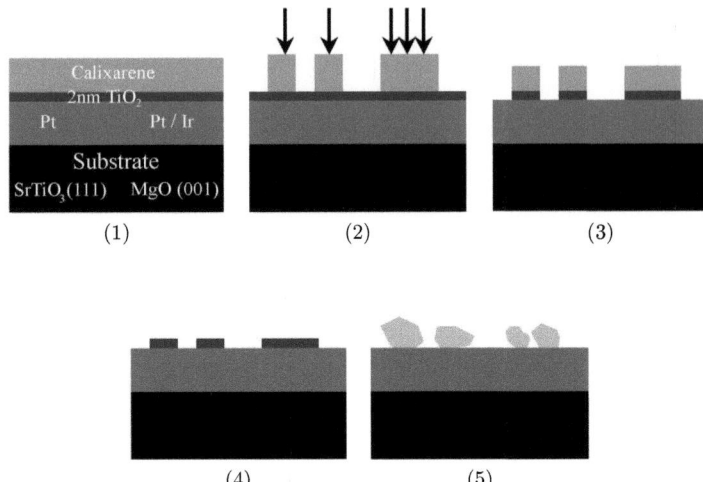

Figure 4.16: Process flow for the additive method using the negative Calixarene resist. (1) Two types of starting substrates: SrTiO$_3$ (111) with epitaxial Pt (111) or MgO (100) with epitaxial Ir/Pt (100) and a Calixarene layer for electron beam lithography, (2) electron beam exposure and development, (3) dry etching, (4) removal of remaining Calixarene in oxygen plasma, (5) Pb(Zr$_{0.4}$,Ti$_{0.6}$)O$_3$ deposition of nominal 20nm. It is supposed that the TiO$_2$ layer is dissolved in the PZT layer. Instead of the negative Calixarene resist, positive PMMA with Cr hard mask and lift-off was used as well (see (2) - (4) of Figure 4.7, p.88).

stable and were removed completely in acetone. At least 20nm Cr was required for lift-off. We used 30nm Cr for our process. This required 100nm of PMMA for a clean lift-off.
Figure 4.17, p.95 shows graphs of the Cr dot diameter as a function of dwell-time for different dot distances. The dot distances were varied from 250nm to 500nm and 1000nm. The PMMA resist layer was 100nm thick, and the thickness of the deposited Cr was 30nm. (a) shows the obtained sizes on 50nm Pt on SrTiO$_3$ (111), and (b) shows the diameters for 50nm Pt/85nm Ir/MgO (100). The difference between the two substrates is the thickness of high density material (50nm Pt in (a), and in (b) 50nm Pt plus 85nm Ir). Enhanced backscattering in expected in (b) leading to a more important proximity effect. In both cases, the electron beam conditions were the same: 20kV and 200pA. On both substrates, the proximity effect leads to an overall larger spot size for dots spaced by only 250nm. The difference is at least 25nm compared to dots which are spaced by 500nm or more. For dots spaced by 250m, above 200μs the proximity effect becomes more important leading to a steep increase of the dot size. In this region, the dots start to touch. On both substrates the dots spaced by 500nm and above, no proximity effect can be observed and the dot size is independent of the spacing. The thickness of the high density materials Pt and Ir influences considerably the dot size. Backscattering is very much enhanced on the substrate having 135nm of high density material below PMMA (b). This leads to overall larger spots on this substrate. Moreover, on this substrate, the Cr dot size decreases linearly

4.4. THE ADDITIVE ROUTE[2]

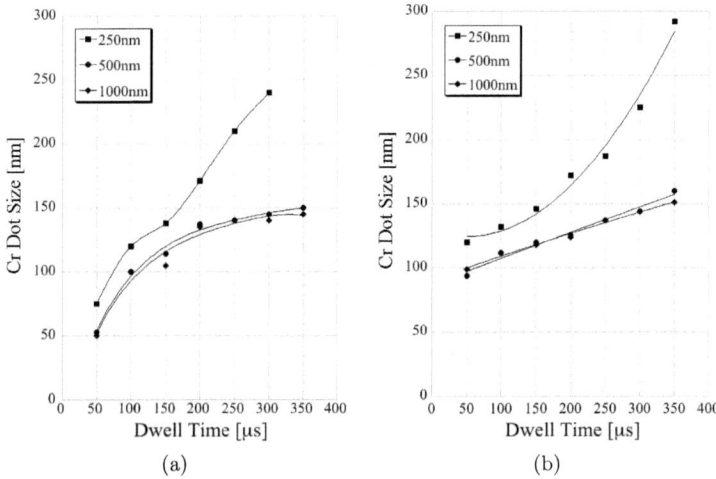

Figure 4.17: Dot diameter of $30nm$ thick Cr dots after lift-off in acetone. The PMMA resist layer was $100nm$ thick. The exposure was done at $20kV$ acceleration voltage and $200pA$ electron beam current. (a) $2nm$ TiO$_2$/$50nm$ Pt (111)/SrTiO$_3$(111), (b) on $2nm$ TiO$_2$/ $50nm$ Pt/$85nm$ Ir/MgO (100).

with the dwell-time for dots with a spacing equal or above $500nm$. On the substrate with a thinner layer of high density material (a), the dot size is decreases considerably below $100\mu s$. This can be interpreted as a decrease in an additional decrease of backscattering with decreasing dwell-time. On this substrate, the dot size reaches a stable value of $150nm$ at $350\mu s$. In the case of thicker high density material below Pt (b) the dot size still increases linearly with dwell-times above $300\mu s$. Figure 4.18, p.96 shows SEM images of holes in PMMA and $30nm$ thick Cr dots obtained after lift-off on $50nm$/SrTiO$_3$ (111) (Figure 4.17(a), p.95) for a center-to-center spacing of $250nm$, $500nm$ and $1000nm$. The large increase in dot size with increasing dwell-time due to an important proximity effect for closely spaced dots contrasts with a less important increase for larger spacings. Figure 4.19, p.97 shows Cr dots obtained after lift-off on the sample with $50nm$ Pt plus $85nm$ Ir below PMMA. It can be seen that the lift-off did not work on the closely spaced dots for dwell-times above $150\mu s$. At $150\mu s$ the PMMA was not removed. The large proximity effect led to rounded PMMA sidewalls after development. For a lift-off process, steep siedwalls are required for the removing agent to access the polymer. Here, the rounded sidewalls were covered by Cr, what in turn made it impossible for the acetone to access the underlying PMMA and to dissolve it. At a dwell-time of $250\mu s$ almost the whole area was developed, what led to an almost complete removal of PMMA during development. Remaining spikes can be seen. These are remaining PMMA hills covered by Cr and also unaccessible to acetone. In contrast, the lift-off worked well on dots spaced by more than $500nm$.

96   CHAPTER 4. FABRICATION OF FERROELECTRIC NANO-STRUCTURES

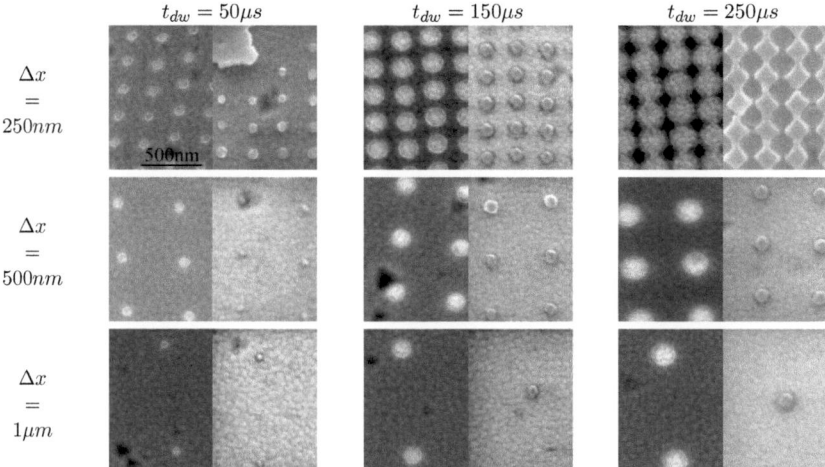

Figure 4.18: SEM images of dots created into $100nm$ thick PMMA at different dwell-times t_{dw} and stepsizes Δx. Here, the step-size equals the distance between the center of neighboring dots. The substrate is $SrTiO_3$ (111) with $50nm$ Pt and $2nm$ TiO_2. Acceleration voltage: $20kV$ with $200pA$. The image on the left shows the image after development (holes), the image on the right shows the Cr dots ($30nm$ thick) obtained after lift-off.

4.4. THE ADDITIVE ROUTE[2]

Figure 4.19: SEM images of dots created into $150nm$ thick PMMA at different dwell-times t_{dw} and stepsizes Δx. The substrate is $2nm$ TiO_2/$45nm$ Pt/$85nm$ Ir/MgO (100). Acceleration voltage: $20kV$ with $240pA$.

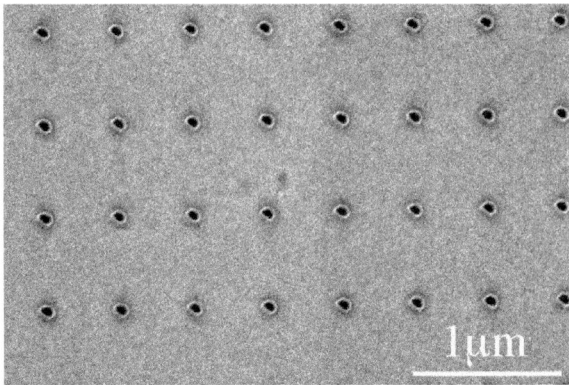

Figure 4.20: Smallest dots obtained with EBL using negative Calixarene resist on $2nm$ $TiO_2/50nm$ $Pt/SrTiO_3$ (111). The dot size is $70nm$. The electron beam conditions were $20kV$ acceleration voltage and $200pA$. The distance between dots is $500nm$ and each dot was written at $4ms$.

4.4.2 Fabrication of TiO_2 Dots using Calixarene[3]

The Calixarene resist was tested on $2nm$ $TiO_2/50nm$ $Pt/SrTiO_3$ (100). The thickness was $60nm$. Exposure took place at two acceleration voltages: $10kV$ and $100pA$, and at $20kV$ $200pA$. The smallest dots had a diameter of $70nm$ and were obtained at the electron beam conditions of $20kV$ and $200pA$, for a spacing of $500nm$ and a dwelltime of $4ms$ (see Figure 4.20, p.98). Much longer dwell-times are necessary to fully develop a Calixarene dot. Using a $20kV$, $200pA$ electron beam and a dot spacing of $500nm$, a dwelltime of $300\mu s$ produces a dot with a diameter of $150nm$ in PMMA. Under the same conditions, $9ms$ are needed to produce the same size in Calixarene. Therefore, the necessary dose to fully develop Calixarene is 30 times higher. At $20kV$ acceleration voltage, the minimum dwelltime required to develop a dot was $1.5ms$. Below, electron beam exposed regions were dissolved during development. The crosslinking of Calixarene under electron beam exposure is a complex process. In PMMA, a chain scission model accounts for the increase of solubility and the observed decrease of the spot size with decreasing exposure time. Calixarene shows a different behavior. At large dwell-times, the spot size decreases when decreasing the dwell-time, passes through minimum and increases again for small dwell-times (see Figure 4.21, p.99). This effect was particularly strong when $10kV$ acceleration voltage and $100pA$ beam current was used. Dots obtained under these conditions and spaced by $500nm$ are shown in SEM images in Figure 4.22, p.100. At the limit of exposure (a) a random dot pattern can be observed showing blurred dots. The regular pattern of dots spaced by $500nm$ is observed at $4ms$ dwell-time (b), but the border of the obtained dots are not well defined. Not well defined borders can be attributed to insufficient development, and appear only for electron beam exposures at $10kV$, $100pA$ acceleration voltage: When the spacing between the dots is increased, backscattering is reduced and unsharp dots appear for higher dwell-times. For $250nm$ spacing, such dots are observed below $3ms$, at $500nm$ spacing below $4ms$, and at $1000nm$ spacing below $4.5ms$. Increasing the dwell-time to 5 and then $6.5ms$ leads

4.4. THE ADDITIVE ROUTE[2]

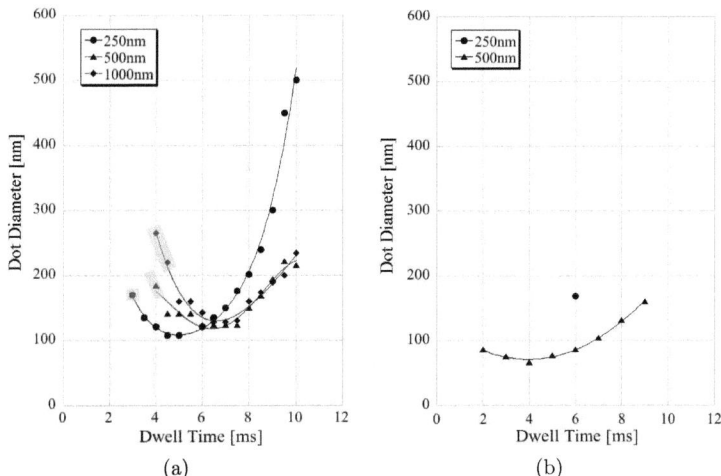

Figure 4.21: Dot diameter as a function of the dwell-time written in $60nm$ thick Calixarene on $2nm$ TiO_2/$50nm$ Pt/$SrTiO_3$ (111). (a) at $10kV$, $100pA$ and (b) at $20kV$, $200pA$. The gray bar indicates dots which show not well defined borders (compare also Figure 4.22(b), p.100).[3]

to well defined dots, and at the same time the dot size is decreased to $123nm$ (d). Increasing further the dwell-time increases the dot size as expected.

Figure 4.21, p.99 shows the dot diameter as a function of the dwell-time and dot spacing for an electron beam of $10kV$ (a) and $20kV$ (b). An interesting observation is the increased spot size for shorter dwell-times. This is evidenced at all conditions by parabolic shaped graphs.[3] As in the case of EBL on PMMA/Pb(Zr,Ti)O_3/$SrTiO_3$, backscattering is reduced for higher acceleration voltages by displacing the interaction volume away from the surface further into the bulk. This leads to overall smaller dot sizes for an acceleration voltage of $20kV$ (b). Here, the smallest dots were obtained at $4ms$ and had a diameter of $70nm$. These dots are shown in Figure 4.20, p.98. The reduction of backscattering also leads to a slower increase of the dot size for long dwell-times, and the parable's shape is wider than in (a). At $20kV$ acceleration voltage, the influence of the increased proximity effect for closely $250nm$ spaced dots is shown by one measurement at $6ms$ (b). The dot diameter of dots spaced by $250nm$ is $180nm$. This is double the value compared to the measured $90nm$ for a spacing of $500nm$.

The influence of the proximity effect is investigated in more detail for exposures at $10kV$ (Figure 4.21(a), p.99).[3] At long dwell-times (above $6ms$), the dot size is considerably higher for closely spaced dots (spacing $250nm$), and moreover, due to the proximity effect, the spot size increases faster for closely spaced dots. At $8ms$ the dot size of $500nm$ spaced dots is $150nm$ and $200nm$ for $250nm$ spaced dots, while at $10ms$ the spot size for closely spaced dots is more than doubled. No proximity effect is observed for dots spaced by $500nm$ and $1000nm$ − they show the same dot size. The most striking feature is the behavior of the dot size below $7ms$. In fact, the smallest dot size is found for exposures with high proximity effect ($250nm$ spaced dots), where

100   CHAPTER 4. FABRICATION OF FERROELECTRIC NANO-STRUCTURES

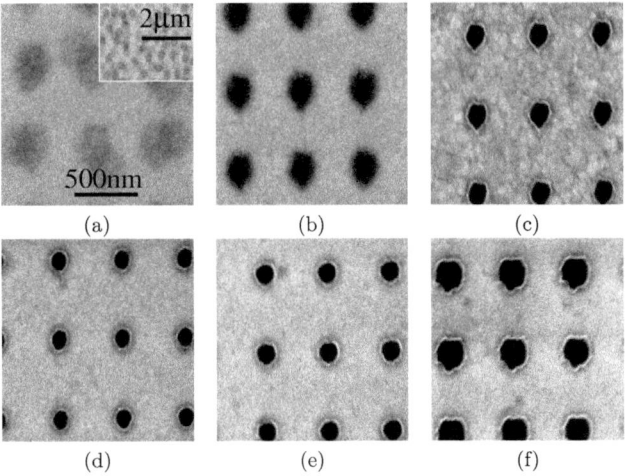

Figure 4.22: SEM images of Calixarene dots after development, written at $10kV$, $100pA$. Different dwell-times have been used: (a) $3.5ms$, (b) $4ms$ ($185nm$), (c) $5ms$ ($150nm$), (d) $6.5ms$ ($123nm$), (e) $8ms$ ($141nm$), (f) $9ms$ ($195nm$).

the smallest dots occur at $5ms$ showing a diameter of $110nm$. The smallest dots for $500nm$ and $1000nm$ spaced dots is found to occur at a longer dwell-time of $7ms$, showing a larger spot size of $125nm$. Striking is not only the increase in spot size with decreasing dwell-time but also the increasing spot size with an increasing spot distance (or reduced proximity effect).[3]
These two behaviors, increasing spot size for decreasing electron doses at low dwell-times, and increasing spot size for increasing electron doses at high dwell-times, can be explained by assuming two chemical mechanisms happening during exposure. Generally, when a polymer is irradiated by an electron beam, two processes are to be taken into account: bond scission and three dimensional reticulation (crosslinking). Bond scission generally leads to higher local dissolution ability whereas three dimensional reticulation accounts for a lower dissolution. In positive resists like PMMA, the former case is dominant, in negative resists like Calixarene the latter. There is a common behavior expected for both types or resist: Bond scission happens at low irradiation times and for three dimensional reticulation, higher electron doses are necessary. Even in the case of PMMA, a three dimensional reticulation is observed for very high exposure times leading to a non dissolvable phase. Bond scission at low irradiation doses is expected to happen in Calixarene as well. As such, the behavior at low irradiation doses can be understood in terms of a competition between bond scission and three dimensional reticulation. Calixarene is very sensitive to electron beam irradiation and partial reticulation takes place even at small doses. Due to the additional backscattering, large dots having blurred borders appear. In this regime, if the electron dose is increased, bond scission starts to counteract this reticulation. This effect is stronger at the border of dots where only partial reticulation takes place. This leads to smaller dots. At the same time, the enhanced dose in the center leads to complete development there and well defined dots with sharp borders are observed. Above the minimum dot size, three dimensional reticulation is the dominant process and larger dots are observed for longer

4.4. THE ADDITIVE ROUTE²

dwell-times. On our substrate, 2.5×10^6 electrons were required to fully develop a dot at an acceleration voltage of $20kV$ and a beam current of $200pA$. This is 25 times higher than the dose found by Fujita et al.,[321] where only 1×10^5 electrons were necessary for the same type of Calixarene. This dose corresponds to the dose for development of PMMA. However, their substrate was Si and the beam condition $50kV$ and $100pA$.

4.4.3 Pb(Zr,Ti)O$_3$ Crystals Obtained on TiO$_2$/Pt/SrTiO$_3$ (111)

Three different types of Pb(Zr$_{0.4}$,Ti$_{0.6}$)O$_3$ depositions were carried out. In the first one, PZT was directly deposited on the prepared substrate. In the other two cases, a PbTiO$_3$ starting layer of 1 or $2nm$ was applied prior to Pb(Zr$_{0.4}$,Ti$_{0.6}$)O$_3$ deposition. The deposition conditions for PZT and PbTiO$_3$ are listed in Table 3.1[2, 3], p.43.

4.4.3.1 Pb(Zr,Ti)O$_3$ Deposition without PbTiO$_3$ Starting Layer

In this deposition type, Pb(Zr$_{0.4}$,Ti$_{0.6}$)O$_3$ with a nominal thickness of $20nm$ was deposited directly onto the Pt surface, which was covered with TiO$_2$ seed dots. Figure 4.23, p.101 shows an SEM image of the obtained crystals. (a) shows the result obtained on small TiO$_2$ seed dots. The dots are around $150nm$ in diameter. As expected, Pb(Zr,Ti)O$_3$ nucleated only on the dots and not elsewhere on the bare Pt. However, on bare Pt, the formation of an additional layer can be observed. The material of this layer is different from the material nucleated on TiO$_2$ seeds. On TiO$_2$ seeds PbO enrichment was found by means of Auger electron mapping (see Figure 4.24(a), p.102). The PZT coverage of a TiO$_2$ was up to four PZT nuclei. Some dots didn't show any nucleus at all. The length of the triangular nuclei's sidewalls varies up to $125nm$. The smallest observed embryos showed a diameter of $20nm$. Figure 4.23(b), p.101 shows an image where Pb(Zr,Ti)O$_3$ nucleated on larger scale TiO$_2$ seed islands. The triangles have a random size distribution. As in the case (a) the triangles are all oriented in the same way with one side-line parallel to the [1$\bar{1}$0] direction of the SrTiO$_3$ (111) substrate. The other side-lines are [0$\bar{1}$1] and [10$\bar{1}$].

Figure 4.23: SEM images showing Pb(Zr,Ti)O$_3$ crystals after the deposition of nominal $20nm$ Pb(Zr$_{0.4}$,Ti$_{0.6}$)O$_3$ directly on the Pt (111) surface with TiO$_2$ seed islands (no PbTiO$_3$ starting layer). (a) Nucleation on TiO$_2$ seed dots, (b) nucleation on lager TiO$_2$ areas. The circles in (b) highlight small triangles grown directly on Pt.

The directions perpendicular to a side-line (in the (111) plane) are [11$\bar{2}$], [$\bar{2}$11], and [1$\bar{2}$1] respectively. The possible side-wall growth facet for the [1$\bar{1}$0] side can be (110), (11$\bar{1}$) or (11$\bar{2}$) planes. The inclination of these planes with the (111) plane are 35°, 70°, and 0° respectively. From a sample where large sized single Pb(Zr,Ti)O$_3$ crystals were obtained on Pt (see Figure 4.25, p.103), it was possible to estimate the side-wall inclination to be rather 70° than 35°. This implies (11$\bar{1}$) type growth facets. For this estimation, combined SEM and AFM measurements have been used.

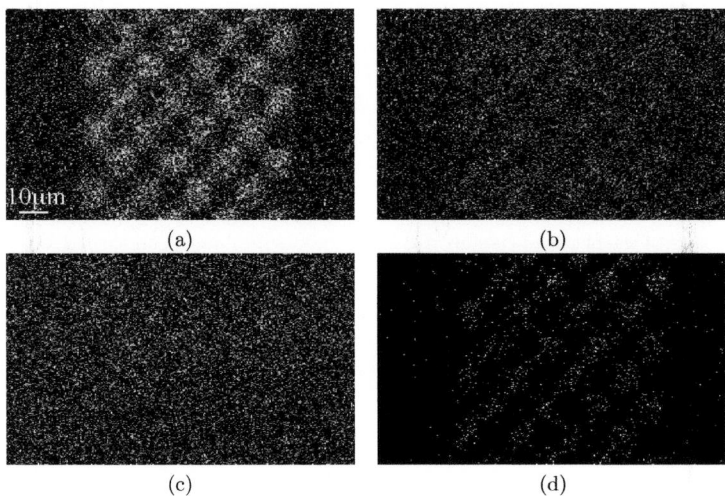

Figure 4.24: Auger surface mapping analysis of Pb (a), Ti (b), Zr (c) and O (d) after the deposition of 20nm Pb(Zr,Ti)O$_3$. No PbTiO$_3$ starting layer was applied prior to PZT deposition. The analyzed area shows a region where 10 × 10μm large TiO$_2$ seed squares have been arranged in a checkerboard.

On bare Pt, the nucleation of a small number of Pb(Zr,Ti)O$_3$ triangular-shaped crystals was observed (see circles in Figure 4.23(b), 101). These triangles were smaller than the ones nucleated on TiO$_2$ and had a lateral size below 50nm. These triangles were still piezo- and ferroelectric, see Figure 5.19, p.134. The surface of the bare Pt changed during the Pb(Zr,Ti)O$_3$ deposition and a thin non-perovskite layer was formed on the Pt. Figure 4.24, p.102 shows a mapping analysis using Auger electron detection. (a) shows the map of Pb. On TiO$_2$ seeds, the contrast of lead is clearly enhanced. Only a slight enhancement of Ti (b) can be seen. The Zr distribution (c) is the same everywhere. (d) shows the contrast of oxygen. On bare Pt, a strong oxygen deficiency is detected. The unknown phase that was formed on bare Pt is a Pb and oxygen deficient layer. PAFM measurements showed that this layer is not piezoelectric. Therefore, this layer could be a pyrochlore phase.

4.4. THE ADDITIVE ROUTE[2]

Figure 4.25: SEM image showing large single crystals of $Pb(Zr_{0.4},Ti_{0.6})O_3$ obtained on bare Pt using a higher lead oxide flow. AFM measurements revealed the height of the crystals a $80nm$.

4.4.3.2 $Pb(Zr,Ti)O_3$ Deposition with 1nm $PbTiO_3$ Starting Layer

Figure 4.26, p.103 shows the results obtained on $50nm$ $Pt/SrTiO_3$ (111). Before the deposition of $Pb(Zr_{0.4},Ti_{0.6})O_3$, a $1nm$ thick $PbTiO_3$ layer was deposited to enhance (001) nucleation (see Table 3.1[2], p.43 for deposition conditions). For $Pb(Zr_{0.4},Ti_{0.6})O_3$ $30nm$ nominal thickness was deposited. (a) compares nucleation on a large area covered with TiO_2 with nucleation on Pt. Very dense nucleation is observed on TiO_2 and a triangular structure can clearly be seen, which indicates (111) growth on TiO_2. On bare Pt, nucleation also took place, but less dense.

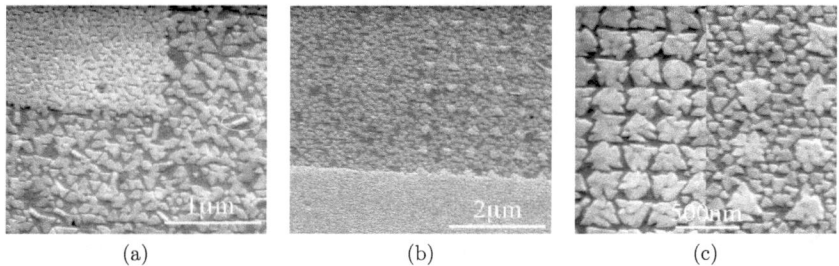

(a) (b) (c)

Figure 4.26: SEM images showing $Pb(Zr,Ti)O_3$ crystals obtained on $TiO_2/Pt/SrTiO_3$ (111). (a) Dense $Pb(Zr,Ti)O_3$ nucleation on large area TiO_2, (b) nucleation on bare Pt, large area TiO_2, and on $500nm$ spaced TiO_2 dots, (c) nucleation on $250nm$ spaced dots compared to $500nm$ dots.

The size of the triangles on bare Pt was observed to be in the range of $20nm$ to $150nm$. The height of the deposited crystals was measured by AFM to be $25nm$. The triangles all had the same orientation with a side-line parallel to the $[1\bar{1}0]$ direction of the $SrTiO_3$ (111). However, rectangular shaped rods can also be observed on bare Pt (indicated by a white circle). (b) compares the nucleation on bare platinum, large area TiO_2 and on $500nm$ spaced TiO_2 dots. The denser nucleation on the dots can clearly be seen. As TiO_2 acts as a sink for PbO molecules, less dense nucleation is expected around the TiO_2 seeds. This can not be observed for $500nm$ spaced dots. (c) compares $250nm$ (left) with $500nm$ (right) spaced dots. The role of TiO_2 seeds as PbO sink can be seen only very near around closely spaced dots. This is observed For TiO_2

dots spaced by 250nm, almost no Pb(Zr,Ti)O$_3$ nucleated between the dots. We also note, that several crystals nucleated on those dots, where up to 7 Pb(Zr,Ti)O$_3$ crystals agglomerated.

4.4.3.3 Pb(Zr,Ti)O$_3$ Deposition with 2nm PbTiO$_3$ Starting Layer

Figure 4.27, p.104 shows Pb(Zr,Ti)O$_3$ nucleation on 500 × 500nm (a, b) and 1.2 × 1.2μm (c) large TiO$_2$ islands. The spacing between the edge was 500nm in (b) and (c), and 1μm in (a). On the smaller TiO$_2$ seed islands (a,b), a non coalescent deposition showing mainly square shaped (001) Pb(Zr,Ti)O$_3$ crystals is observed. Sporadic triangular shaped crystals are also found on the TiO$_2$ seeds. Their size is similar to the size of the square shaped crystals. Coalescence took place on the large area seed TiO$_2$ island (c) where isolated square shaped crystals can only be distinguished at the border but not in the center. On bare Pt, nucleation of small triangles with a size about 20nm (a) and large square shaped crystals took place. The growth of square shaped crystals on Pt start from a circular shaped structure (highlighted in (a)). The square shaped crystals on the bare Pt and on the TiO$_2$ seeds show a uniform size distribution with a lateral size of 100nm. One sideline of the squares is preferentially oriented in the [1$\bar{1}$0] direction with respect to the SrTiO$_3$ substrate (see indication in (b)). Otherwise, the squares are randomly oriented.

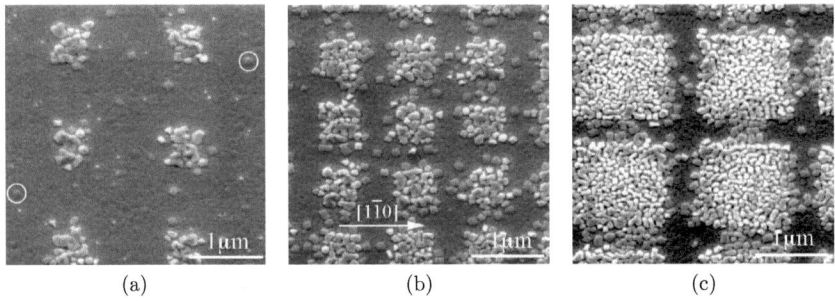

Figure 4.27: SEM images showing Pb(Zr,Ti)O$_3$ growth on TiO$_2$/50nm Pt/SrTiO$_3$ (111) using a 2nm thick PbTiO$_3$ layer before Pb(Zr$_{0.4}$,Ti$_{0.6}$)O$_3$ deposition. (a) growth on 500 × 500nm^2 large TiO$_2$ seeds with an edge to edge spacing of 500nm, (b) on 500 × 500nm^2 large TiO$_2$ seeds with an edge to edge spacing of 1000nm, (c) 1.2 × 1.2μm^2 large areas spaced by 500nm. The white circles in (a) highlight the initial shape of square shaped squares.

The preferential orientation could arise from stress in the substrate which has a dimension of 5 × 10mm^2. The small triangles in (a) could not be observed between the closely spaced TiO$_2$ seeds in (b,c). Here, the TiO$_2$ seeds work as a sink for PbO species creating a PbO deficient region around the seed which inhibits nucleation of small triangles there. Figure 4.28, p.105 shows the depletion region appearing in the vicinity of large scale TiO$_2$ seeds when they are spaced more than 1μm. Almost no nucleation of small triangles can be observed within a border of 1μm around the TiO$_2$ seed. This is coherent with the appearance of some small triangles for TiO$_2$ seeds spaced by 1μm (Figure 4.27(a), 104). For larger spacings Figure 4.28(a), p.105 the PbO sink is not felt anymore by the nucleation process and a constant density of small triangles appear. The nucleation density of triangles on bare Pt far away from TiO$_2$ seeds was measured

4.4. THE ADDITIVE ROUTE[2]

to be 60 times smaller than on the TiO_2 seeds. On small TiO_2 seed dots, the shape of the PZT crystals was changed from triangular to square-shaped. This is due to the high Pb containing, Zr-less deposition. Figure 4.29, p.105 shows an SEM image of the obtained crystals. These square shaped PZT crystals show a uniform size of less than $200nm$ side-length, and exactly one crystal grew on one TiO_2 seed.

Figure 4.28: SEM images showing a depletion of $Pb(Zr,Ti)O_3$ nucleation on bare Pt around TiO_2 seed areas. (a) $5 \times 5\mu m^2$ seeds spaced by $5\mu m$, (b) zoom onto a $2 \times 2\mu m^2$ large feature.

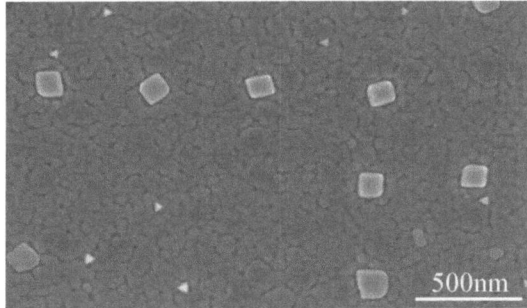

Figure 4.29: SEM images showing square shaped $Pb(Zr,Ti)O_3$ crystals nucleated on single dots.

4.4.4 $Pb(Zr,Ti)O_3$ Crystals Obtained on TiO_2/Pt/Ir/MgO (100)

For the deposition of $Pb(Zr,Ti)O_3$ on the sample with Pt (100), no $PbTiO_3$ seed layer was deposited before $Pb(Zr,Ti)O_3$ deposition. $Pb(Zr_{0.4},Ti_{0.6})O_3$ was deposited with a nominal thickness of $20nm$. Figure 4.30(a), p.107 shows the depositions obtained on bare Pt and on a region where TiO_2 dots have been defined beforehand with a spacing of $1\mu m$. The nucleation of an unknown phase was evidenced by SEM. A dense nucleation can be observed on bare Pt, while for the region containing some amount of TiO_2, the density of nucleated species is reduced (a). These white dots always nucleated on the bare Pt and not on TiO_2 seed islands. The observed white dots arise from two types of nucleated material (b). The first type (1 in Figure 4.30(b). p.107) is large, has often a rhombohedral shape and is up to $70nm$ in height. The other type (2) is smaller, embedded in the also existent underlayer, and appears less well defined. Type 1 appears preferentially in the vicinity of the border of TiO_2 seed areas. Figure 4.30(c), p.107 shows part of the $5\mu m$ large TiO_2 frame and the patterned region (upper right). The large dots grow preferentially in near the frame. It is noted that on the frame, no white dots are observed. Nucleation controlled mechanisms play an important role also for depositions on Pt (100). Around TiO_2 seeds, a depletion layer can be observed, where no white spots exist at all. This region spreads over $500nm$ as it can be observed on large TiO_2 seed areas (d). Again, no dots are observed on the TiO_2 seeds. White spots are only seen on bare Pt. As seen in (a), the presence of TiO_2 reduces the density of their growth. Figure 4.30(e), p.107 shows a region where TiO_2 seed islands of $500 \times 500nm^2$ spaced by $1\mu m$ have been defined. Between the squares, white spots exist. Their density is $1.3 dots/\mu m^2$. (f) shows the same pattern, but separated by $500nm$. The white dots can only be observed outside the patterned area. Between the square shaped TiO_2 islands, the their growth was inhibited. The density on bare Pt without TiO_2 seeds was measured to be $2.3 dots/\mu m^2$.

The nucleation of $Pb(Zr,Ti)O_3$ was only observed on TiO_2 seeds. Figure 4.31(a), p.108 shows randomly nucleated $Pb(Zr,Ti)O_3$ material (in white circles) on a large scale TiO_2 seed feature. Piezoelectric sensitive AFM measurements showed that this material is piezoelectric. The shape of nucleated $Pb(Zr,Ti)O_3$ material on large scale TiO_2 seeds is random. (b) shows a region where TiO_2 dots have been created. The dots have a diameter of $150nm$ and are spaced by $500nm$. Most of the dots do not show preferential nucleation of PZT. Others clearly show square shaped features which are all oriented in the same direction. One edge of the features is parallel to the (100) direction of the MgO substrate (see (d) for the direction). This indicated (001) growth of the crystals. (c) shows the topography of the TiO_2 dots without nuclei from the $Pb(Zr,Ti)O_3$ deposition have the same morphology than the deposition on bare Pt. The white spots show the same morphology as well, however they show a brighter contrast due to an other composition. The square shaped look of nucleated dots is clearly be seen in SEM images in (d). The [100] is indicated by an arrow which is parallel to an edge of the nuclei. The features have an average size of $200nm$ and are made of several nuclei.

4.4.5 Site Controlled Nucleation and Growth

The fabrication process used in the additive route relies on nucleation controlled issues. It was shown, that the Pt (111) is not a good candidate for $Pb(Zr,Ti)O_3$ nucleation, and that nucleation started only on TiO_2 seeds. The results were most striking on samples when no $PbTiO_3$ layer was applied: Almost no $Pb(Zr,Ti)O_3$ nucleated on bare Pt. However, when a $PbTiO_3$ starting layer was applied, or when using a Pt (100) surface, nucleation also took place on bare Pt, but at a rate which was 60 times less. In these cases, typical nucleation related features could be observed which highlight the importance of surface energies, diffusion on surfaces and the PbO flux. As such, a depletion layer of nucleation around large scale seeds was observed on Pt (111) samples; the additional PbO flux was captured almost entirely by the TiO_2 seeds leading to square shaped crystals there and to triangular shaped crystals on the depleted bare Pt. In what follows, we perform calculations to explain the phenomena of the depletion layer. We refer to the sample where $Pb(Zr,Ti)O_3$ was deposited on Pt (111) using a $PbTiO_3$ starting layer of $2nm$.

4.4. THE ADDITIVE ROUTE[2]

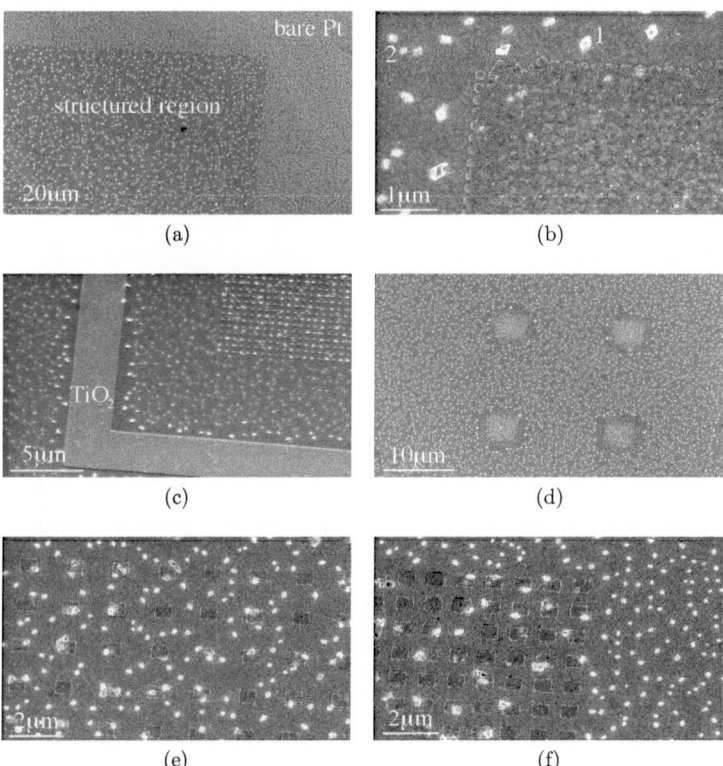

Figure 4.30: Site controlled nucleation on TiO$_2$/Pt (100)/Ir/MgO (100). No PbTiO$_3$ starting layer was deposited prior to PZT deposition. SEM images showing nucleation on large scale TiO$_2$ seed islands on Pt (100). Two type of outgrows: (1) rhombohedral shaped, (2) undefined shape.

CHAPTER 4. FABRICATION OF FERROELECTRIC NANO-STRUCTURES

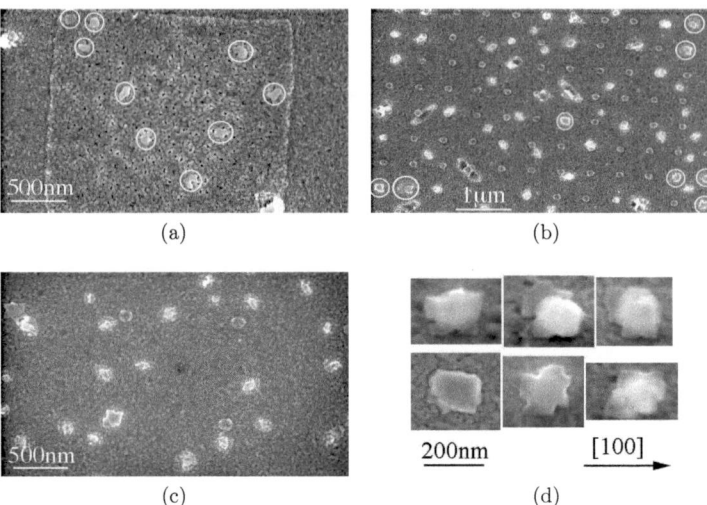

Figure 4.31: Site controlled nucleation on TiO_2/Pt (100)/Ir/MgO (100). No $PbTiO_3$ starting layer was deposited prior to PZT deposition. SEM images showing nucleation on small scale TiO_2 seed islands on Pt (100).

In our work we apply a kind of local seed layer to increase the density of nucleation sites as well as to decrease the critical radius and thus the critical activation energy. Although TiO_2 does not exhibit the same crystalline structure as PZT, the thin layer effectively reduces the nucleation activation energy for the condensation of the perovskite phase. The main reason is the more exothermic reaction of TiO_2 and PbO to form $PbTiO_3$ than the one of ZrO_2 and PbO to form $PbZrO_3$. This is made evident by the fact that the PbO vapor pressure in the PZT system increases with the Zr content.[345] A further important feature of $Pb(Zr,Ti)O_3$ growth is the much larger PbO pressure above PZT as compared to the ones of B-site atoms or their oxides. At the PZT film growth temperatures (570°C), condensation of PbO is not stable. PbO can be considered as the most diffusing, most volatile species during PZT growth. This allows to a certain extent the application of the classical nucleation theory for monomers.[141] Essential for perovskite formation is a large enough density of PbO molecules. Ti and Zr can be considered as (more slowly migrating) adsorption sites. On a bare Pt (111) surface all PbO desorbs at the perovskite growth temperature.[57] PbO that missed to attach to a B-site species desorbs again. Obviously, the seed captures a part of the PbO flux arriving near its borders. The width of the depletion zone is related to the diffusion distance λ of the PbO species along a given direction during its residence time on the substrate. It is approximately given by $\lambda = \exp(\frac{E_a-E_d}{2kT})$. E_a is the activation energy for ad- or desorption, and E_d the one for diffusion (see also the corresponding section on p.23). The frequency constants (attempt rates or inverse hopping times) for diffusion ν_d and desorption ν_0 are supposed to be equal (which is in fact a reasonable assumption[141]). For the $1\mu m$ depletion zone as observed in Figure 4.28, p.105, and $T = 873K$ deposition temperature, we obtain $E_a - E_d = 1.2eV$. This value is compatible with an estimation of the adsorption energy E_a according to Cockcroft's argument saying that the arrival rate on

4.4. THE ADDITIVE ROUTE

	TiO_2	Pt
E_d	$0.8eV$	$0.7eV$
E_a	-	$2.0eV$
τ_r	-	$0.33s$
τ_{cs}	$0.01s$	-

Table 4.5: Table of the calculated parameters for the nucleation model.

a given site must be larger than the desorption rate in order to obtain a film. This statement translates into the equation $\Phi \tau_r > 1$ if we express the arrival rate Φ in monolyers per second (τ_r is the residence time). Experimentally we found that the critical flux for PbO deposition at $500°C$ amounts to about $2.5ML/s$ (static deposition). The corresponding adsorption energy is found to be $1.9eV$, the diffusion energy thus would amount to about $0.7eV$.

In order to model the observed behavior for the first few seconds of the deposition, we assume that the PbO species are the only ones that are responsible for the diffusion on the micrometer scale. TiO diffusion is of the order of at best $50nm$, as judged form the density of PbTiO$_3$ nuclei.[57] The adatom density n_a at point (x,y) obeys a rate equation of the following form:

$$\frac{dn_a(x,y)}{dt} = \Phi(x,y) - R_{des}(x,y) + R_{dif}(x,y) - R_{cs}(x,y)$$

where Φ is the deposition flux density, R_{des} the desorption rate, R_{dif} the diffusion rate and R_{cs} the chemisorption rate (PbO that is incorporated into stable, solid phase). The desorption rate is proportional to the adatom density, and inversely proportional to the residence time τ_r:

$$R_{des}(x,y) = \frac{n_a(x,y)}{\tau_r} \quad \text{and} \quad \tau_r = \nu_0^{-1} \exp \frac{E_a}{kT_s}$$

The diffusion rate follows from the standard diffusion equation $R_{dif}(x,y) = D\Delta n_a(x,y)$, where the diffusion constant D depends on the activation energy E_a for hopping to the next adsorption site, the attempt rate ν_d, and the hoping distance a_0 corresponding to the periodicity of the Pt (111) surface $a_{Pt}/\sqrt{2}$.

$$D = \frac{3}{4} a_0^2 \nu_d \exp \frac{-E_d}{kT}$$

The chemisoprtion rate has the same structure as the desorption rate. We just introduce a mean time τ_{cs} between arrival and chemisorption to the titania film. Considering a circular seed, the steady state diffusion equations are:

$$D\left(\frac{\partial^2 n_a}{\partial r^2} + \frac{1}{r}\frac{\partial n_a}{\partial r}\right) - \frac{n_a(x,y)}{\tau} + \Phi = 0 \quad \text{and} \quad \frac{1}{\tau} = \frac{1}{\tau_r} + \frac{1}{\tau_{cs}}$$

The solutions are modified Bessel and modified Hankel functions for inside and outside the seed respectively.

Outside seed:

$$n_a = \Phi \tau_s \left(1 - \gamma_s \frac{K_0 q_s r}{K_0 q_s a}\right)$$

Inside seed:

$$n_a = \Phi \tau_p \left(1 - \gamma_p \frac{I_0 q_p r}{I_0 q_p a}\right)$$

Here, τ is the overall residence time. The index $_s$ denominates the situation on the TiO$_2$ seed, and $_p$ on the bare Pt. a is the distance from the center of the seed, r is the radius of the seed (equal to $100nm$). Using the continuity condition at the border of the seed, the relevant parameters can be calculated. The values are given in Table 4.5, p.109.
Using these parameters, the stationary solution of n_a is calculated as depicted in Figure 4.32, p.111. The capture rate is given in units of B-site species arrival rate (which is $1ML/s$ perovskite), the adatom density n_a in monolayers (perovskite). The values have been chosen as to describe the situation with PbTiO$_3$ starting layer. The equilibrium vapor pressure of PbO ($p_{e,PbO}$) exceeds by far the equilibrium vapor pressures of titania and zirconia species. We can estimate the Gibbs' free-energy difference ΔG per PbO molecule in the perovskite lattice according to

$$\Delta G \approx -kT \ln \frac{p_{PbO}}{p_{e,PbO}}$$

Using the equilibrium pressures by Härdtl et al.[345] ΔG of Pb(Zr$_0.5$Ti$_0.5$O$_3$) is $0.09eV$, less exothermic than the one of PbTiO$_3$. Depending on the actual PbO vapor pressure during deposition, the activation energy for nucleation may exhibit substantial differences. The PbO vapor pressure is the essential parameter determining activation energies and critical dimensions of nuclei. The monomer approximation using PbO vapor and adatom density gives at least a good qualitative description of the nucleation in the PZT system. The result shows that more PbO arrives by diffusion than by direct impingement. That's in fact alike the case in classic nucleation theory.[345] The smaller the seed the more diffusion becomes important as compared to direct impingement from the gas phase. The seed gathers enough PbO to nucleate the perovskite out of the seed layer. Outside the seed layer, there is always an excess of B-site species, as the PbO has a much shorter residence time. A value of $0.34s$ is calculated with the values listed in Table 4.5, p.109. Therefore, after 1 or $2nm$ of deposition, when embryonic nuclei may reach the critical size, there is a lead deficiency with respect to the stoichiometry. The model parameters yield a stoichiometry Pb/Ti of 0.7 after $1s$ of deposition. Even if pyrochlore or fluorite is nucleating, PbO desorption will continue to a much larger extent on these phases as compared to the perovskite phase. In this way it is possible to get films with pyrochlore as dominating phase on Pt, even so the same process yields perovskite on the titania seed. The interesting question is why the titania seed layer works, as there is more Ti there right from the beginning. The answer is that the titania seed layer chemisorbs well the lead, and since much more lead oxide arrives at the beginning with PbTiO$_3$, stoichiometry is reached after a given time. The amount of material deposited in total must now fit to the critical nucleus size. The thickness of seed layer has thus to be tuned to obtain a good timing. It is indeed observed that the titania seed layer is optimal for $2-3nm$. Given the flat shape of the nuclei, the critical radius must be of the same order of size.

4.4. THE ADDITIVE ROUTE[2]

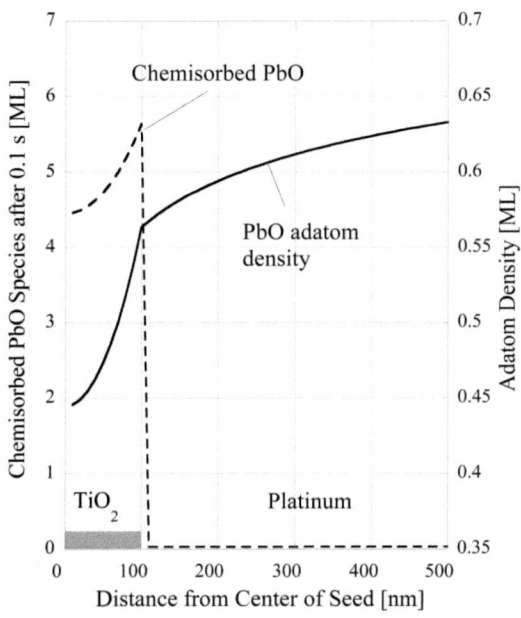

Figure 4.32: Model of chemisorption and adatom density in the vicinity of a TiO_2 seed.

4.5 Summary and Conclusion

In this Chapter, the fabrication of small ferroelectric structures on single crystalline substrates by means of electron lithography was demonstrated using subtractive and additive approaches. The resolution limit of the electron beam process using PMMA was found to be $70nm$. The main limiting factor is the backscattering of electrons into the resist film when working on films of high density materials such as Ir, Pt, but also PZT. This especially the case when working at intermediate acceleration voltages around $20kV$. Backscattering was also the limiting factor for resolution when using the promising negative resist Calixarene. Extremely high doses were necessary to develop this resist (40 times higher than for PMMA). It showed an interesting development behavior at small doses where an increase in the exposure time led to smaller dots. This could be explained by increased bond scission which counteracts the reticulation in this regime. Above a critical value, only reticulation takes place and the "normal" behavior (larger dots with longer exposure times) was observed.

Using the subtractive route, electron beam lithography with PMMA and Cr hard masking produced PZT features down to $70nm$. Below $150nm$, the edges of the features were blurred out due to the loss of resolution at these sizes.

The patterning of the $2nm$ thick TiO_2 used in the additive route was successfully demonstrated using PMMA and Calixarene resists. An additional difficulty arose with respect to resolution: A slight loss of resolution was observed due to the high density metallic layers below. On patterned TiO_2 on Pt (111), the additive method worked out well: During $Pb(Zr_{0.4},Ti_{0.6})O_3$ deposition, triangular crystals almost exclusively nucleated only on the TiO_2 seeds and not elsewhere. The next Chapter shows that these crystals are ferroelectric. The triangles are all oriented in the same way, which indicates epitaxial (111) growth. However, an thin layer of unknown material was also deposited on bare Pt. Auger electron measurements revealed that this layer is heavily oxygen and lead deficient. This is in contrast to the material on the TiO_2 seeds which shows strong PbO enrichment. Surprisingly, no $Pb(Zr,Ti)O_3$ nucleated on Pt steps. In an attempt to switch the orientation from (111) to (100), a $PbTiO_3$ starting layer of $1nm$ was applied prior to $Pb(Zr_{0.4},Ti_{0.6})O_3$ deposition. The $Pb(Zr,Ti)O_3$ crystals still showed the triangular (111) orientation on the TiO_2 seeds, and dense nucleation of triangles also took place on bare Pt. Using a $2nm$ thick starting layer of $PbTiO_3$ prior to $Pb(Zr_{0.4},Ti_{0.6})O_3$ deposition was able to switch the triangular shape to square shaped (100) features. Dense nucleation was observed on TiO_2 seeds, but some triangles also nucleated on bare Pt. The nucleation density on TiO_2 was 60 times higher than on bare Pt. The effect of surface diffusion could well be observed. On bare Pt around large scale TiO_2 seed islands, a depletion region is seen where almost no triangles nucleated. Here, in the vicinity of the TiO_2 seed, the lead deficiency is so large that no nucleation could take place. A nucleation model was developed, which was able to calculate deposition parameters of nucleation, such as activation energies of adsorption and diffusion. The calculated depletion layer fitted well our observations.

On patterned TiO_2 on Pt (100) the influence of TiO_2 seeds on the nucleation could also be observed. No $PbTiO_3$ starting layer was applied prior to $Pb(Zr_{0.4},Ti_{0.6})O_3$ deposition. On bare Pt dense nucleation of small features was observed. The features have an undefined shape and are not piezoelectric (see next Chapter). In the vicinity of TiO_2 seeds the nucleation was less dense but larger spots appeared. No nucleation of spots was observed on the TiO_2 seeds. However, on the TiO_2 seeds a third phase could be observed at some places. On large scale TiO_2 seeds, these features were randomly shaped, but on the small TiO_2 dots, square shaped features were obtained. The next Chapter shows that these features are ferroelectric.

Chapter 5

Piezoelectric Atomic Force Microscopy (PAFM) Measurements

The ferroelectric behavior of the features obtained by subtractive and additive routes were characterized by means of piezoelectric atomic force microscopy (PAFM). This technique uses a conductive AFM tip which is scanned in contact with the surface of the ferroelectric thin film or crystal. The tip works as a movable top electrode. An AC modulation is applied to the tip, and the bottom electrode is grounded. This induces the converse piezoelectric effect in the ferroelectric layer, which is sandwiched between the bottom electrode and the tip. This local piezoelectric vibration of the ferroelectric is transferred mechanically to the tip and to the cantilever. To control the cantilever's position (or the force set by the user), our system uses a laser beam, which is reflected on the cantilever to a photosensitive detector. While low frequency movements of the cantilever arising from the sample's topography are immediately corrected by a feed-back loop (in combination with the scanner which adjusts the sample height), in order to maintain a constant force on the cantilever, the high frequency signal of piezoelectric vibration (we use a signal at $17kHz$) is transmitted unfiltered to the photosensitive detector. This signal is also called the "error signal" or "$A-B$" signal. It is fed to a lock-in amplifier which compares it to the initial excitation AC modulation. From this comparison, the lock-in amplifier provides two output values: the phase shift of the vibration with respect to the excitation signal measures the direction of the polarization (up or down), and the amplitude measures the extent of local piezoelectric vibration. Before showing the results, a short introduction to PAFM is first given. Then, it is shown how we calibrated the tip vibration to be able to convert the voltage signal from the lock-in into the real tip vibration. The results are first shown for the subtractive route. Here, the characterization of epitaxial continuous $Pb(Zr_{0.4},Ti_{0.6})O_3$ films on Nb-doped $SrTiO_3$ (100) substrates are given. It is shown how a continuous film can be poled locally, and how the poling voltages, time and spacing between two points influence the obtained dot sizes. Then, it is shown, that local electrical training by a conductive AFM tip induced changes in the domain configuration, what led to instantaneous high piezoelectric vibration. Then, the results obtained on the patterned features are given. Finally, the results for patterns fabricated by the additive route are shown. Here, $Pb(Zr_{0.4},Ti_{0.6})O_3$ was deposited on surfaces of epitaxial Pt (111) and Pt (100) covered with $2nm$ thick seed sites. First, results for features obtained without a starting layer of $PbTiO_3$ are presented: for $Pb(Zr_{0.4},Ti_{0.6})O_3/2nm$ $TiO_2/Pt/SrTiO_3$ (111) and on $2nm$ $TiO_2/Pt/MgO$ (100). Then, the obtained results for PZT depositions on $2nm$ $TiO_2/Pt/SrTiO_3$ (111) with $1nm$ and $2nm$ thick $PbTiO_3$ starting layers are presented.

5.1 Introduction to PAFM

The development of the first scanning tunneling microscope in 1982 by Binnig and Rohrer[346] was the starting point. This technique soon developed into a whole family of scanning probe microscopes (SPM).
Scanning tunneling microscopy used in combination with ferroelectric materials was probably used for the first time to relate surface roughness of a polyimide layer to the alignment of the subsequent ferroelectric liquid crystal.[347,348] The alignment of ferroelectric liquid crystals were observed on graphite with scanning tunneling micrscopy. This work did not investigate ferroelectric properties. SPM techniques applied to ferroelectrics include the measurement of the capacitance,[349] electrostatic force,[350] the nonlinear dielectric microscopy,[351] and the piezoelectric vibration under an applied AC field (PAFM). Long range electrostatic tip-surface interactions can be measured as well.[352,353] Variations in the electric stray field above 180° domains are sensed in electrostatic force microscopy in a non contact mode.[354] The electrostatic interaction between the tip/cantilever and the bottom electrode of the sample may play a significant role in the formation of the PAFM contrast.[353,355] The properties of the cantilever may strongly influence the measurements.[355] The electrostatic force can strongly interact with soft cantilevers. In recent years, scanning probe microscopy (SPM) based techniques have been successfully employed in the characterization of ferroelectric surfaces on the micron an nanometer levels[356–358] via the piezoelectric effect. Lateral and vertical movements of the cantilever can be detected. This allows to distinguish between in plane a-domains and out of plane c-domains. Such investigations have been done on $BaTiO_3$ single crystals.[354] The lateral resolution of $20nm$ suggested a penetration depth of the electric field of the same order.[354] Small domains in single crystal $BaTiO_3$ of sizes below $100nm$ were achieved only for very high fields and short pulses $< 30\mu s$. A density of $9Gbit/cm^2$ was observed, however backswitching occurred after some days.[354]
Local domain writing of a continuous ferroelectric layer by means of an AFM tip could be used as a memory.[354,359] For the AFM tip/$Pb(Zr,Ti)O_3$/Pt sequence, there is substantial spatial variation in the electric field,[359] and not simply $E = V/d$. The coercive field in the inhomogeneous electric field distribution determines the bit size. The coercive field was found to be dependent on the pulse width (large pulse width - small coercive field). On $270nm$ thick $Pb(Zr_{0.4},Ti_{0.6})O_3$ (111) on $Pt/SiO_2/Si$ it was found that the bit size is linearly dependent on the pulse voltage and the logarithmic value of the pulse with. The linear dependence of the bit size on the logarithmic value of pulse width was explained from the relationship between the switching time and electric field.[360]
Tybell et al.[81] investigated ultrathin sputtered, (001) oriented $PbZr_{0.2}Ti_{0.2}O_3$ films on conductive, Nb-doped $SrTiO_3$ which showed ferroelectricity down to a thickness of $4nm$. The poled regions were stable for at least $180h$. For measurements on such thin films, PAFM is an excellent method because it circumvents problems related to leakage. Piezoelectric measurements as well as electrostatic measurements (EFM, to measure polarization charges) were performed to investigate the stability of switched regions. EFM measurements were performed to test the surface polarization charges. They decreased to zero over time, showing the screening of appearing compensation charges at ambient conditions. In order to observe the limit of stable polarization in $PbTiO_3$, Roelofs et al.[59] deposited a very thin film of a chemical solution containing the precursors of $PbTiO_3$. After a heat treatment, randomly distributed $PbTiO_3$ grains formed on Pt. The grain-size was in the range of $20 - 200nm$. PAFM studies on relatively thick films showed equidistant stripes of $12 - 80nm$ within the grains. Using both, the tip's lateral and the vertical vibration for the analysis, the stripes could be identified as 90° domains.
Chen et al.[361] observed 90° domain formation after switching a c-axis oriented, $20 - 200nm$

thick Pb(Zr$_{0.2}$Ti$_{0.8}$)O$_3$ film on SrTiO$_3$ (100) with $50nm$ La$_{0.5}$Sr$_{0.5}$CoO$_3$ bottom electrode. The formation of 90° domains has two competing consequences: The electrostatic energy increases, but the elastic energy decreases. If the elastic energy decrease exceeds the increase in electrostatic energy, the formation of 90° domains is favorable.

Luo et al.[64] fabricated PbZr$_{0.52}$Ti$_{0.48}$O$_3$ and BaTiO$_3$ nanotubes exhibiting piezoelectric hysteresis and ferroelectric properties. AFM measurements revealed a d_{33} of $100pm/V$. The loop showed a lower d_{33} value after reversing the DC.

Eng et al.[362] used contact, non-contact and friction force mode to decode topographic and ferroelectric contributions of cleaved tri-glycine sulphate (TGS) (010) surface. No additional AC signal was applied to the tip. Measurements in the non-contact mode revealed the inverse surface polarization, and lenticular shaped domains have been observed. Those domains disappeared when heating above the Curie temperature of $49°C$. In the contact mode, the lenticular domains could not be detected. It was concluded, that the detection of those domains in the non-contact mode was possible due to surface polarization which interacted with the effective spring constant of the cantilever. Friction force was shown to depend on the sign of polarization at the surface.

Gruverman et al.[363] used a gold-coated Si$_3$N$_4$ cantilever to image the piezoresponse of ferroelectric Pb(Zr,Ti)O$_3$ surfaces. The lower limit is defined as the estimated contact diameter of the tip. Switching of domains as small as $80nm$ was performed on polycrystalline PZT. Fatigue measurements (scanning for $1h$ above coercive field) showed unswitchable regions after fatiguing. Ganpule et al.[223] investigated the 3D arrangement of domains in $400nm$ thick PbZr$_{0.2}$Ti$_{0.8}$O$_3$ films on conductive Nb-doped SrTiO$_3$ (100). A Pt/Ir tip with a radius of $20nm$ was used. The tilting of a-domains leads to a small polarization in the normal direction. Applying a electric field leads to shear deformation characterized by the piezocoefficient d_{15}. This can be sensed by a torsion of the cantilever. The direction of the torsion gives an indication of the polarization's direction. It was therefore possible to confirm the head-to-tail configuration across 90° domains. Switching of 180° domains led to an unstable state which reversed after some time. The nucleation of reversed domains was observed to occur along 90° domains. Ganpule et al.[364] used Si tips coated with nickel silicide for electrical conduction on PbZr$_{0.2}$Ti$_{0.8}$O$_3$/LSCO/SrTiO$_3$ (100). The radius was $5-15nm$. TEM showed the a-domain structure (see Figure 2.12, p.35) and PAFM observation revealed the same. Switching did not change the a-domain structure. 90° domains act as centers for polarization reversal of 180° domains.

5.2 Quantifying the PAFM Piezoelectric Response

From the PAFM measurement, the lock-in amplifier gives us a phase and a voltage signal. The phase allows us to conclude whether the polarization is upwards or downwards. The voltage amplitude is proportional to the local piezoelectric expansion of the film. In order to quantify the measurements, we need to calibrate two values:

- the real amplitude of tip vibration for an applied AC signal
- the piezoelectric coefficient of the ferroelectric layer

To assert the real tip vibration amplitude, a calibration was done using a $600nm$ thick epitaxial Pb(Zr,Ti)O$_3$ film on conductive Nb-doped SrTiO$_3$ (100). Large scale capacitors were defined by $300 \times 300 \mu m^2$ Pt top electrodes. Ferroelectric d_{33} loops have been measured on these capacitors at $17kHz$ by an interferometer (see Figure 5.1(a, gray line), 116). For calibration, the same loop was also acquired by an AFM tip immobilized on the top electrode. The excitation signal

116 CHAPTER 5. PIEZOELECTRIC ATOMIC FORCE MICROSCOPY (PAFM) MEASUREMENTS

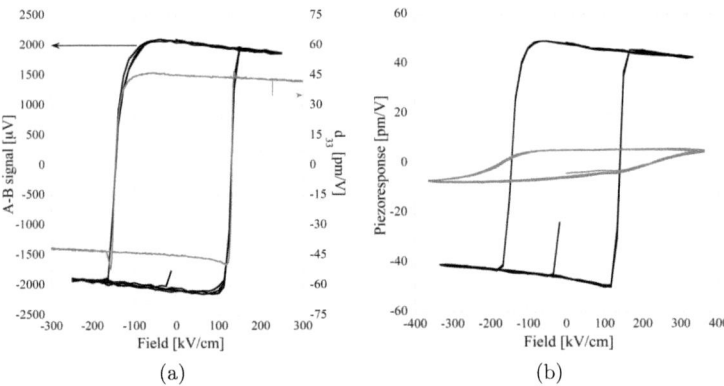

Figure 5.1: Loops acquired on a 600nm thick Pb(Zr$_{0.4}$,Ti$_{0.6}$)O$_3$ film on SrTiO$_3$ (100). (a) In order to quantify the real tip vibration, d_{33} loops have been acquired on a large scale top electrode by an interferometer (gray line), and by an AFM tip immobilized on the same top electrode (black line). The calibration factor was found as $32\frac{pm_{amp}}{mV_{A-B,rms}}$. (b) In order to estimate the real d_{33} as measured by a tip in contact with the PZT film, a loop was acquired using an AFM with the tip immobilized on the large scale top electrode (black line), and then directly on the PZT film (gray line). The conversion factor between the two was 7.5. Now, using the result from (a), a d_{33} value for local piezoelectric vibration can be calculated.

frequency was the same as used in the measurement by the interferometer, *i.e.* $17kHz$. The tip was short-circuited with the top electrode to avoid parasitic effects, and the top electrode was contacted to the excitation signal from the lock-in. Figure 5.1(a, black line), p.116 shows the obtained $A-B$ signal in μV. Comparing the $A-B$ signal to the loop from the interferometer, the calibration factor is

$$32\frac{pm_{amp}}{mV_{A-B,rms}}$$

The locally induced vibration is dependent on the local piezoelectric coefficient and the amplitude of the excitation signal. Dividing the real tip vibration by the excitation signal amplitude gives a pseudo d_{33} value for the local piezoelectric coefficient. However, this value measured locally by the tip would be much lower than the true d_{33} measured on large scale top electrodes by the interferometer. This is because the experimental set-up of a tip on a thin film cannot be regarded as a parallel plate configuration. The electric field below the tip is highly non-uniform. It is concentrated in a region just below the tip and decreases rapidly radially.[365] One can even assume that the field emanating from the tip does not penetrate the whole thickness of the ferroelectric layer, and that only several tens of nanometers are affected.[365] Moreover, other unknown factors affect the measurement, such as the conductivity of the very thin conductive layer on the tip and a thin H$_2$O layer present at the surface, which might lead to parasitic capacitances. Figure 5.1(b), p.116 shows a comparison of a loop measured on top large scale top electrodes (black line), and measured locally, using the tip as top electrode (grey line). The vibration using the tip as top electrode is about 7.5 times lower than measured using the large scale the top electrode. Using this conversion factor of 7.5 only gives a rough estimation

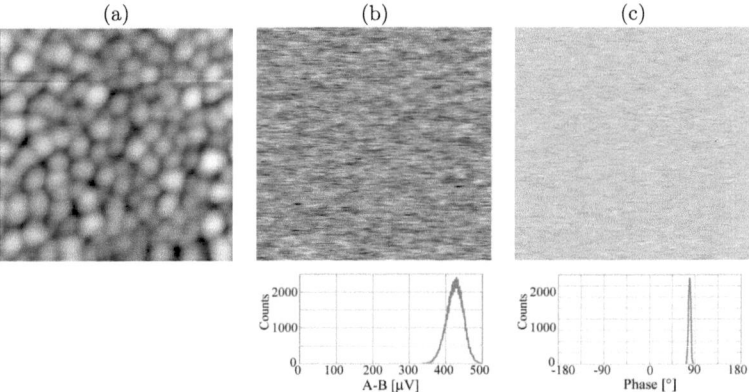

Figure 5.2: $1 \times 1 \mu m^2$ PAFM scan on $100 nm$ as deposited Pb(Zr$_{0.4}$,Ti$_{0.6}$)O$_3$ film deposited on SrTiO$_3$ doped with $1\% - at$. Nb. (a) topography, (b) amplitude of piezoelectric vibration, (c) phase shift with respect to the excitation signal ($2V_{rms}$). $+90°$ corresponds to up polaritzation, $-90°$ is the inverse direction. Beneath (b) and (c) are shown a line scan and the histogram of the corresponding image.

of a d_{33} value of the film, and has thus not been used in this work for the interpretation of results. Instead, a pseudo d_{33} value is used in ferroelectric loop measurements. It was found that nevertheless, the real vibration of the tip was proportional to the excitation voltage. The pseudo d_{33} value was calculated by dividing the real tip vibration by the amplitude of the excitation voltage. This allows us to compare measurements acquired at different AC voltages. In the figures, we refer to this value as to the "piezoresponse".

5.3 PAFM on Structures Obtained by the Subtractive Route

5.3.1 Measurements on the Continuous Pb(Zr,Ti)O$_3$ Film

On the continuous Pb(Zr,Ti)O$_3$ films, PAFM was used to measure the piezoelectric response, to perform local poling, and local training of the film.

5.3.1.1 PAFM Polarization Experiments[4]

Figure 5.2, p.117 shows a $1 \times 1 \mu m^2$ PAFM scan of an as deposited $100 nm$ thick continuous Pb(Zr$_{0.4}$,Ti$_{0.6}$)O$_3$ film on SrTiO$_3$ (100) doped with $1\% - wt$. Nb, which reflects the typical piezoelectric response of a Pb(Zr,Ti)O$_3$ film. (a) shows the topography, (b) amplitude, and (c) the phase image. Below the amplitude and phase images, histograms show the distribution of the measured vibration amplitude from the photodetector ($A - B$ signal) and the phase. The scan was performed at an excitation signal of $17 kHz$ and $2V_{rms}$. The average piezoelectric response from the histogram is $425 \mu V_{A-B,rms}$. Using the before calculated calibration factor of

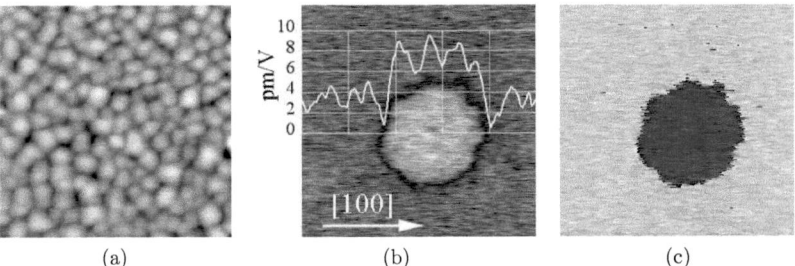

Figure 5.3: $1 \times 1 \mu m^2$ scan on a $200nm$ thick Pb(Zr$_{0.4}$,Ti$_{0.6}$)O$_3$ film on Nb-doped SrTiO$_3$ (100) of a region which was poled at $+30V$ to the tip. The excitation signal was 0.4V rms. (a) topography, (b) amplitude of vibration, (c) phase. The left side of the loop corresponds to the up polarization. An upward shift of the loop can be observed reflecting the higher intensity of the down polarization in (b).

$32 \frac{pm_{amp}}{mV_{A-B,rms}}$, the average piezoelectric response corresponds to a piezoelectric vibration amplitude of $13.6pm$. Considering the $2V_{rms}$ excitation signal, the average d_{33} relative to the tip is $d_{33} = 4.8pm/V$. This is a typical value found on continuous Pb(Zr$_{0.4}$,Ti$_{0.6}$)O$_3$ films regardless of the thickness of the film, tip, and excitation voltage. In Figure 5.1(b, gray loop), p.116 a value of $4.5pm/V$ can be seen for a $600nm$ thick Pb(Zr$_{0.4}$,Ti$_{0.6}$)O$_3$ film on SrTiO$_3$ (100). The phase of the as deposited film in Figure 5.2, p.117 is uniform. From the histogram, a narrow distribution is deduced with an average value at a phase shift of around 90°. No counts were detected at $-90°$. This 90° phase shift corresponds to a polarization which is directed from bottom to top. The value of the phase shows a 15° shift with respect to the 90°. The same shift is detected for down polarizations and can be attributed to parasitic capacitances. The difference between down and up polarizations is 180°.

Figure 5.3, p.118 shows a typical image, which is obtained when poling locally the Pb(Zr,Ti)O$_3$. It shows a $1 \times 1 \mu m^2$ scan with topography (a), amplitude (b) and phase image (c) of a $200nm$ thick Pb(Zr$_{0.4}$,Ti$_{0.6}$)O$_3$ film on SrTiO$_3$ (100). Before the measurement, the tip was immobilized in the center, and the film was locally poled by applying $+30V$ to the tip during $1s$. The polarization changed locally from up to down (dark contrast in the phase image (c)). The obtained dot is about $400nm$ in diameter and not perfectly circular as expected. The shape is sometimes delimited by lines which are parallel to [100] directions (b). This is an indication that the epitaxial nature of the film influence the poling. The amplitude is higher in the opposite polarization direction. The inset in (b) shows the piezoelectric response over a line scan. It can be seen, that the polarization can be doubled in the opposite direction. This would correspond to a shift in the hysteresis loop upwards. At zero field, this loop shows a d_{33} value of $4.5pm/V$ in the up polarization (negative field values), and $6.4pm/V$ in the opposite direction (positive field values). It was generally easy to switch the polarization from up to down but more difficult to pole it again in the other direction. The size of a poled dot depended on the applied voltage and on the pulse width (see Figure 5.4, p.119).

5.3. PAFM ON STRUCTURES OBTAINED BY THE SUBTRACTIVE ROUTE

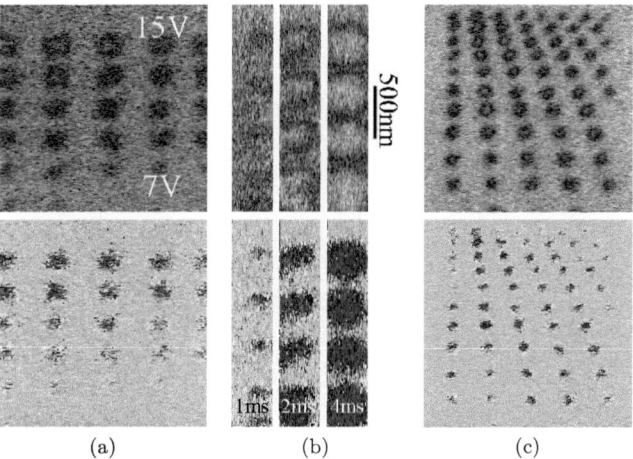

Figure 5.4: $2 \times 2\mu m^2$ PAFM scan on $200nm$ as deposited $Pb(Zr_{0.4},Ti_{0.6})O_3$ film on $SrTiO_3$ (100) doped with $1\% - at$. Nb. Top: amplitude images, bottom: phase images. (a) variable pulse voltages to the tip from $+7V$ to $+15V$ at a pulse width of $60ms$. (b) pulses written at $+30V$ and $1ms$ (left), $2ms$ (middle), and $4ms$ (right). (c) constant pulse parameters ($+13V$, $50ms$) but variable separation distances: horizontally from $300nm$ (down right) to $100nm$ and vertically from $250nm$ to $50nm$.

5.3.1.2 Nano-Training and Recovery of Ferroelectricity[5]

The possibility to recover ferroelectric properties by applying a pulse well above the coercive field is reported by a few authors. Depinning of domains is often cited to be the origin of the recovery.[17,41] Restoring of the ferroelectric loop's squareness was observed by Scott et al.[17] in polycrystalline KNO_3 thin films. A specimen was fatigued to 20% of its initial spontaneous polarization. A $30\mu m$ pulse of $15V$, much larger than the coercive field, could nearly fully recover the spontaneous polarization. On other electrodes, the application of a pulse worsened the polarization.

Restoring of properties by applying an activation pulse was also observed on $SrBi_2Ta_2O_9$ (SBT) integrated capacitors on $Si/SiO_2/Ti/Pt$. The polarization charge was markedly increased on $2.7 \times 2.7\mu m$ capacitors.[22] Fatigue could also be restored, but not fully, on large $Pb(Zr,Ti)O_3$ capacitors.[33] The result indicates that applying large voltage can partly move the pinned domains in the fatigued film. Bellur et al.[41] compared epitaxial and polycrystalline $PbZr_{0.53}Ti_{0.47}O_3$ films. The loop of an epitaxial $Pb(Zr,Ti)O_3$ film grown on Pt/MgO (100) could be recovered after having been fatigued by applying a pulse of $8-10V$ during $2s$. This could not be obtained on polycrystalline films indicating that domain wall pinning is more pronounced in this case.

Fatigue refreshment was investigated on $600nm$ thick $(Pb, La)TiO_3$ films on Pt/MgO.[366] After application of a very long DC voltage of $-20V$ to the top electrode for $10^4 s$ to the as-deposited sample, the remanent polarization was more than doubled. Fatigue during 10^7 cycles at $12V$ was performed what decreased the polarization by 25%. Applying again $-20V$ for $10^4 s$ restored the polarization loop and increased slightly the coercive voltage. Fatiguing directly the as-deposited sample and applying $20V$ afterwards increased the polarization also by a factor of more than two. It was concluded that this film had many defects and that the application of a DC voltage afffected defect dipoles and domain reconstructing.

Ramesh et al.[367] found that the in a $PbZr_{0.2}Ti_{0.8}O_3/YBa_2Cu_3O_7$ heterostructure fatigue process is reversible by application of a 2 second pulse at the fatigue voltage.

Gruverman et al.[39] performed attempts for rejuvenation withan AFM on structured $200nm$ thick Ca-, Sr-, and La-doped $Pb(Zr_{0.4},Ti_{0.6})O_3$ capacitors with an area of $1 \times 1.5 \mu m^2$. The AC training for rejuvenation was done at $4V_{rms}$, $500Hz$, well above the coercive field. The training worsened the situation: Much larger regions showed a small amplitude. Subsequent scanning revealed the same initial situation suggesting again strong negative imprint.

In our training experiments, we used an immobilized AFM tip on a continuous $200nm$ thick $Pb(Zr_{0.4},Ti_{0.6})O_3$ film on $SrTiO_3$ (100) doped with $1\%wt$. Nb. For the loops, the DC voltage was applied from $-20V$ to $+20V$. The same spot was then submitted to a series of pulsing experiments, where Figure 5.5, p.121 shows a selection in sequential order. The pulses were applied at a voltage of $\pm 20V$ and $30ms$. The pulses were applied just before the loop measurement. The indication 50 pulses means 25 times a cycle of $\pm 20V$. (a) shows the loop of the as deposited film. The remanent d_{33} values are in the typical range observed on continuous films and show a shift downwards ($d_{33} = +3pm/V$ and $-4.1pm/V$). The saturation d_{33} is symmetric ($\pm 5pm/V$). Between (a) and (b) two pulse experiments were done at 50 pulses before the loop measurement in(b). (b) shows the loop after the third application of 50 pulses (total 150 pulses). It can be seen that time is an important factor. One loop measurement lasts for $6min$ and consists of 4 consecutive loops. Just after the pulsing, the d_{33} value is increased to the double value ($-11pm/V$ at $-20V$) compared to the value of the as deposited film. After the first switching, $d_{33} = +13pm/V$ at $+20V$. However, switching again leads to a d_{33} which is decreased $-8pm/V$ at $-10V$ with respect to the previous one, and the response decreases towards the maximum DC voltage of $-20V$. It ends at a value close to the d_{33} of the as deposited film. Going again to the positive DC values leads to an observable hump in the loop (gray area). The decrease in d_{33} might be due to pinning of domains in this range. Switching again to the positive DC region leads to the appearance of another but much smaller hump (gray zone) in the positive DC region. Finally, the values stabilize at the same values measured on the as deposited film. (c) shows the same loop measurement, but after a total of 250 pulses. The increase in the total amount of pulses increased the sensibility of the film to the training. The same training as before (50 pulses) now leads to an overall larger increase in d_{33} initially. The loop became more rectangular which indicates an increased importance of $180°$ switching. The pinning-humps on the negative DC side are much more pronounced as before and exist for all partial loops on the negative side. After 4 loops, the d_{33} value is still higher than the ones of the as deposited film. (d) shows the same measurement after a total of 400 pulses. The rectangular shape of the loop can clearly be seen. The pinning-humps still exist (and only on the negative DC side) but are less pronounced than in (c). The initial d_{33} was larger then $20pm/V$. After 4 loops (or $6min$), the d_{33} value is decreased but still amounts to $15pm/V$. Interestingly, the rectangular loops in (c) and (d) show a similar shape than the loops measured by interferometry on $600nm$ thick $Pb(Zr_{0.4},Ti_{0.6})O_3$ on $SrTiO_3$ (100) in Figure 3.12(a), p.54 featuring sharp switching on the positive DC voltage and rounded shape on the negative side. (e) shows loops without preliminary pulsing, measured $10min$ after completion of (d). The d_{33} values are now greatly decreased. However, the d_{33} values at remanence are now higher ($+5.4pm/V$ and $-5.8pm/V$ at remanence and $\pm 8pm/V$ at saturation) compared to the as deposited film (see above). (f) shows the loops obtained directly after application of 500 pulses (950 total pulses). Now, the inverse behavior is observed: Initially, the measured piezoresponse is smaller and stabilizes at a larger value of $20pm/V$. This configuration was stable for at least 2 hours.

5.3. PAFM ON STRUCTURES OBTAINED BY THE SUBTRACTIVE ROUTE

Theory shows that the increase in d_{33} is related to a dramatic change in the domain configuration and that it corresponds well with the sum of contributions from the clamped film ($105 pm/V$) plus the contribution of mobile a-domains ($60 pm/V$).[5]

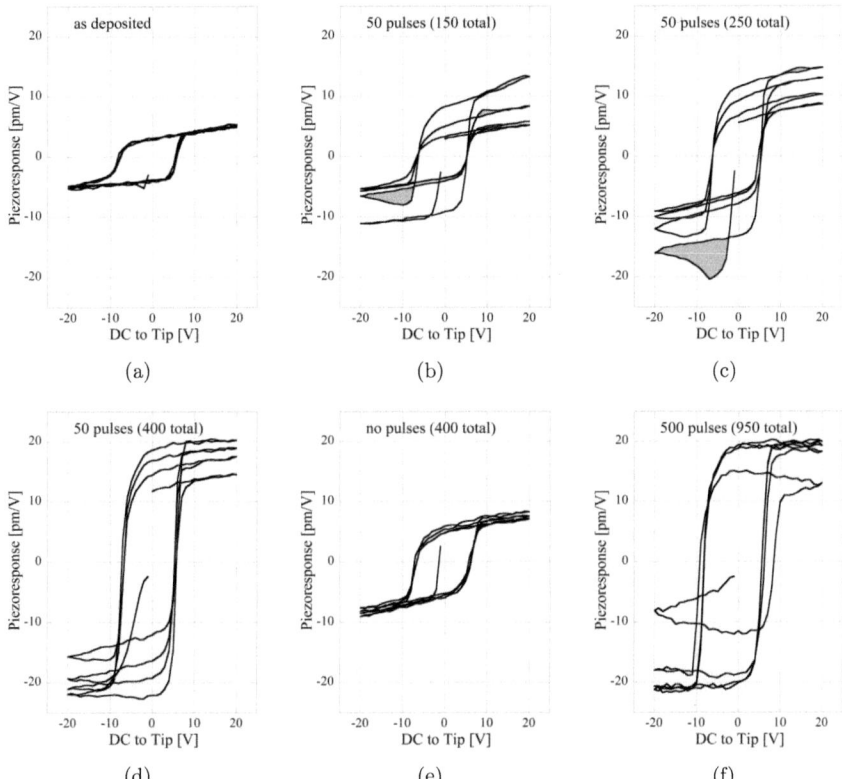

Figure 5.5: Electrical nano-training: Selected sequence of loops measured on a single point by an AFM tip immobilized on an epitaxial $200 nm$ thick $Pb(Zr_{0.4},Ti_{0.6})O_3$ film on $SrTiO_3$ (100). For some loops, a pulse program with pulses of $\pm 20V$, $30ms$ was applied just before the loop measurement. The indication 50 pulses means 25 times $\pm 20V$.[5]

122 CHAPTER 5. PIEZOELECTRIC ATOMIC FORCE MICROSCOPY (PAFM) MEASUREMENTS

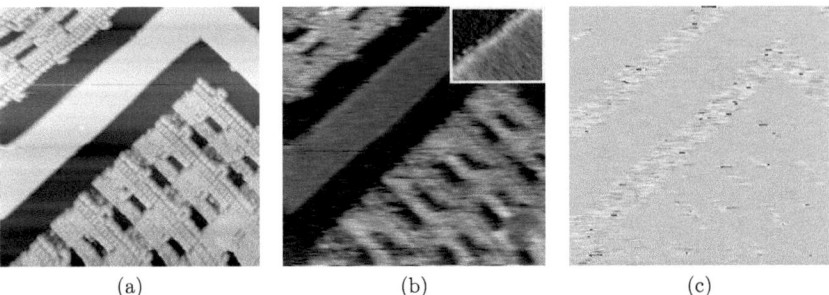

(a)   (b)   (c)

Figure 5.6: $20 \times 20 \mu m^2$ scan on a $200nm$ thick $Pb(Zr_{0.4},Ti_{0.6})O_3$ film on $0.2\% - at.$ Nb-doped $SrTiO_3$ (100) after structuring and Cr removal. The image was acquired at $1V_{rms}$, $17kHz$. (a) topography, (b) amplitude image, (c) phase image. The inset in (b) shows a zoom on the frame. The amplitude of the frame is about 3 times less than in the structured regions. The response on $SrTiO_3$ is zero, no phase can be detected there.

5.3.2 PAFM Measurements on Patterned Features

Figure 5.6, p.122 shows a $20 \times 20 \mu m^2$ PAFM scan on the structures obtained after dry etching and lift-off. The starting point was a $200nm$ thick $Pb(Zr_{0.4},Ti_{0.6})O_3$ film on $SrTiO_3$ (100). After the dry etching process, the height of the obtained features is $200nm$ indicating complete removal of $Pb(Zr,Ti)O_3$. (a) shows the topography featuring the patterned regions with small ferroelectric items (down right) and a large scale frame. (b) shows the amplitude image of the PAFM scan and (b) the phase. From (b) it can be seen, that the polarization is still up. (a) shows zero piezoresponse on attacked regions indicating that all $Pb(Zr,Ti)O_3$ was removed there. In those regions, the phase is not detectable and shows random values (c). It can clearly be seen that in the region of patterned small $Pb(Zr,Ti)O_3$ features, the amplitude is much higher than on large scale $Pb(Zr,Ti)O_3$ (frame, quasi-thin film). A slight increase of the piezoresponse (b) can also be observed on the left border of the frame (see inset in (b)). Figure 5.7, p.123 shows three zoom-ins into the patterned regions of Figure 5.6, p.122 where the amplitude images are on top and the phase images at the bottom. (a) is a $5 \times 5 \mu m^2$ scan, (b) and (c) are $2.5 \times 2.5 \mu m^2$ scans. a1 to c7 indicate different regions where a certain type of feature was patterned. Regions with smaller features show a higher piezoelectric response than regions with features having larger lateral sizes. The average piezoresponse was measured. For squares which were designed at $100 \times 100nm$ dimensions, a d_{33} of above $20pm/V$ was measured, while circles with a diameter of $300nm$ show the same response of the thin film ($3 - 4pm/V$). The average values of each region can be found in the figure caption of Figure 5.7, p.123. Table 5.1, p.124 lists the measured average d_{33} as a function of increasing piezoresponse for different shapes. It can be seen that the length of the smallest lateral size determines the piezoresponse. Therefore, the bars with the lateral dimensions of $900 \times 150nm^2$ show larger response ($4.5pm/V$) than circles with a diameter of $300nm$ ($3.5pm/V$). Shrinking the bar to a rectangular shape of $150 \times 150nm^2$ further increases the response up to $17pm/V$. For the triangle with a side length of $200nm$, one could expect a similar response as for the $150 \times 150nm^2$ squares. However, the electron beam lithography process led to blurred out corners what in turn led to an increase of the dimension.

5.3. PAFM ON STRUCTURES OBTAINED BY THE SUBTRACTIVE ROUTE 123

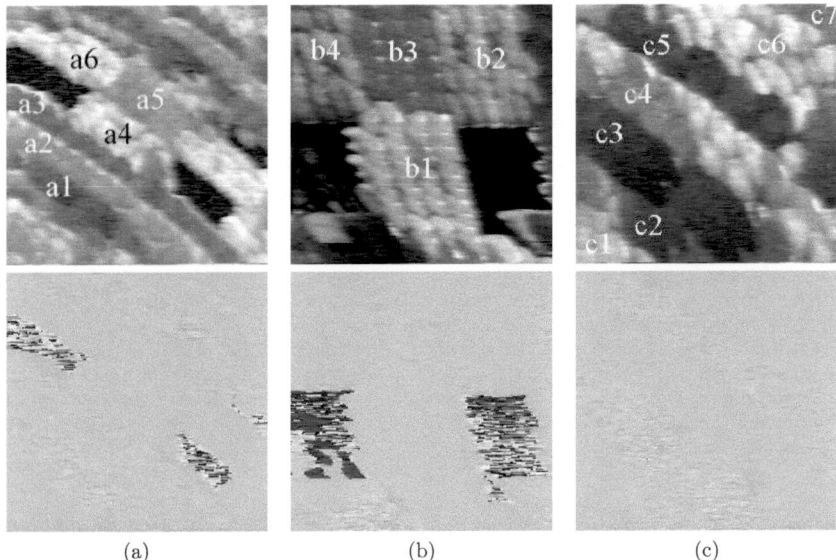

Figure 5.7: PAFM images on differently patterned regions of a $200nm$ thick $Pb(Zr_{0.4},Ti_{0.6})O_3$ film on $1\% - wt.$ Nb-doped $SrTiO_3$ (100). Top: Amplitude images, bottom: Phase images. The scans have been acquired at zero DC filed and at an AC of $1V_{rms}$ and $17kHz$. (a) $5 \times 5\mu m^2$, (b,c) $2.5 \times 2.5\mu m^2$ scans. a1-c7 indicate regions with a certain feature-type. Depending on the feature-type, the average piezoresponse varies as follows. a1: $200 \times 1000nm^2$ $4pm/V$, a2: triangles ($200nm$ lateral size) $9pm/V$, a3: circles ($300nm$ diameter) $3.5pm/V$, a4: $150nm$ squares $19pm/V$, a5: $200nm$ squares $9pm/V$, a6: $100nm$ squares $24pm/V$; b1: $100nm$ squares $19pm/V$, b2: $150nm$ squares $14pm/V$, b3: $200nm$ squares $8pm/V$, b4: $100 \times 300nm^2$ squares $12pm/V$; c1: $100 \times 300nm^2$ features $15pm/V$, c2: $150 \times 900nm^2$ $4.5pm/V$, c3: $200 \times 1000nm^2$ $3pm/V$, c4: triangles ($200nm$ lateral size) $10pm/V$, c5: circles ($300nm$ diameter) $3pm/V$, c6: $100 \times 100nm^2$ $20pm/V$, c7: $200 \times 200nm^2$ $8pm/V$.

124 CHAPTER 5. PIEZOELECTRIC ATOMIC FORCE MICROSCOPY (PAFM) MEASUREMENTS

Shape	Dimension	Area $[10^4 \mu m^2]$	Piezoresponse pm/V
▬	$1000 \times 200 nm^2$	2000	3.5
●	$300 nm$	706	3.5
▬	$900 \times 150 nm^2$	1350	4.5
■	$200 nm$	400	8.5
◣	$200 nm$	200	10
▬	$300 \times 100 nm^2$	300	13.5
■	$150 nm$	225	16.5
■	$100 nm$	100	22

Table 5.1: Average piezoresponse of different shapes as a function of the lateral size, fabricated from a $200nm$ thick Pb(Zr$_{0.4}$,Ti$_{0.6}$)O$_3$ film on SrTiO$_3$ (100).

The effect of the lateral size is seen comparing the $300 \times 100 nm^2$ bar ($13.5 pm/V$) with the circle of $300 nm$ diameter $3.5 pm/V$.
Loop measurements on individual features showed the same behavior. Figure 5.8(a), p.125 shows loop measurements performed on single structures with a height of $200 nm$. The enhancement in the piezoelectric response when the film is confined laterally can clearly be observed. On the smallest feature, a hump between 0 and $10V$ can be seen on both sides of the polarization. This arises due to contributions from 180° domains. One factor leading to an increased piezoresponse upon lateral confinement is un-clamping . The equation[368] to calculate the piezoelectric coefficient $d_{33,f}$ of a clamped film is

$$d_{33,f} = d_{33} - \frac{2s_{13}^E}{s_{11}^E + s_{12}^E} d_{31}$$

where s is the compliance of Pb(Zr,Ti)O$_3$, d_{33} the real unclamped piezoelectric coefficient, and un-clamping is accounted for by d_{31}. The term in front of d_{31} can be estimated[368] to -1. For Pb(Zr,Ti)O$_3$, d_{31} is roughly $-\frac{1}{2} d_{33}$, and hence un-clamping of the film can at maximum double the $d_{33,f}$. In our case, we observe an increased piezoelectric coefficient which is up to 6 times larger than the value of the film. Another contribution to the piezoelectric coefficient might arise from domain wall motions (unpinning) and removal of 90° domains. This explanation is supported by the findings from training experiments on continuous Pb(Zr,Ti)O$_3$ films (p.121) where an increase in the piezoresponse up to $20 pm/V$ was found. Figure 5.8(b), p.125 shows the increase of the piezoresponse as a function of the feature's aspect ratio for square shaped features. The effect of un-clamping and domain contributions are indicated. An estimated 'real' d_{33} is shown using the conversion factor of 7.5 between the local PAFM response and the continuous thin film.
The increase of piezoresponse with decreasing feature size was reproduced on a sample with $150 nm$ thick Pb(Zr$_{0.4}$,Ti$_{0.6}$)O$_3$ on SrTiO$_3$ (100). Loop measurements performed on features with different sizes showed the same dependence (see Figure 5.10, p.127). Figure 5.9, p.126 shows the switching of a single feature in an array of a features with $150 nm$ lateral size. (a) the as dry etched features show uniform up polarization. Again, an enhanced piezoresponse is observed at the border. As the features are spaced only by $200 nm$ (edge-to-edge), the pattern show a rectangular shape due to the large AFM tip radius. One feature was switched at $+15V$ to the tip, which is well above the coercive voltage. The PAFM scan after switching is shown in

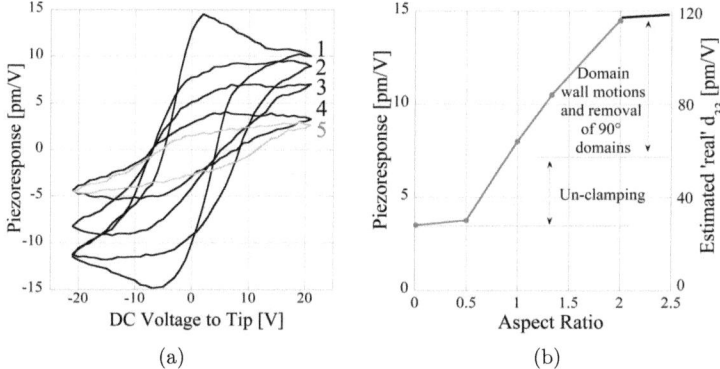

Figure 5.8: Feature size dependence of ferroelectric loops measured by means of PAFM at $17kHz$, $0.4V_{rms}$. The $Pb(Zr_{0.4},Ti_{0.6})O_3$ was $200nm$ thick. Nb-doped $SrTiO_3$ (100) served as a substrate. (a) Loops measured on features with lateral sizes of $100nm$ (1), $150nm$ (2), $200nm$ (3), $400nm$ (4), and on the continuous film (5). (b) the increase of the as a function the aspect ratio of the feature.

(b). The phase image shows complete switching of the feature. The amplitude of the switched feature is less than in the opposite direction. (d) shows a zoom onto the switched feature in (b). The reduced amplitude is well seen. Moreover, the patterns show a certain contrast which is the same for each feature. This can be attributed to the geometry of the tip. Switching the top neighboring feature also switched the two next neighbors. Figure 5.11, p.127 shows the switching of a $100 \times 300nm^2$ feature (a), a triangular feature (b) and a $1000 \times 200nm^2$ (c) feature. The switching voltage was again $+15V$ to the tip. In (a) the right side of the bar was switched completely, the left part didn't. A zig-zag shaped border between the switched and non-switched region can be seen. Hence, no domino effect exist which would lead to the switching of the whole feature. The sharp nature of the border indicates 180° domains on both sides. The triangular shaped feature in (b) was again switched completely. Partial switching occurred on the $1000 \times 200nm^2$ bar. The observed uniform up polarization was not observed for all features after dry etching. Figure 5.12, p.128 shows features obtained by writing single dots during the electron beam lithography. Both scans (top: amplitude image, bottom: phase image) are $400 \times 400nm^2$ and are acquired as dry etched and after lift-off. (a) shows a feature obtained from a $70nm$ dot, and (b) from a $130nm$ dot. Dry etching was not complete in this case, and a $50nm$ thick $Pb(Zr,Ti)O_3$ layer remains. It is noted, that the response of the remaining layer is not uniform and dark (zero response) spots exist. This can be due to the dry etching process which deteriorates the ferroelectric properties. The polarization in the center of the feature is inverse with respect to a $30nm$ large outer ring. This ring shows a high piezoresponse in the top part and a low response with the tendency to switch into the opposite direction in the lower part. A possible explanation could be the influence of d_{15} when the tip touches the side-wall. The proposed domain pattern along the dotted line for the feature in (b) is schematically shown in (c). It features an a domain to account for the almost zero amplitude in the bottom right edge and a 180° domain on the other end. Domain patterns were also observed on large scale patterns. Figure 5.13, p.129 shows features fabricated from a $200nm$ thick $Pb(Zr_{0.4},Ti_{0.6})O_3$

126 CHAPTER 5. PIEZOELECTRIC ATOMIC FORCE MICROSCOPY (PAFM) MEASUREMENTS

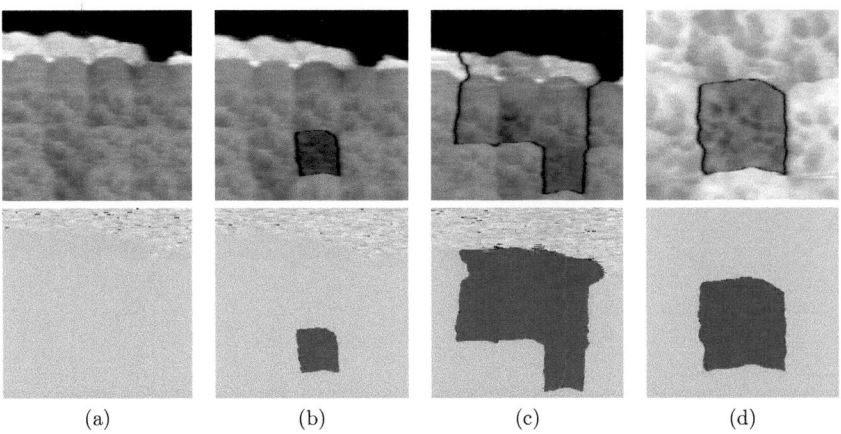

(a)   (b)   (c)   (d)

Figure 5.9: Switching experiment: Amplitude (top) and phase (bottom) PAFM images of $150nm$ features fabricated from a $200nm$ thick $Pb(Zr_{0.4},Ti_{0.6})O_3$ on $SrTiO_3$ (100). (a) $1 \times 1\mu m^2$ scan before switching, (b) $1 \times 1\mu m^2$ scan after switching at $+15V$ to tip, (c) after switching the top neighboring feature, (d) $500 \times 500nm^2$ detail scan of (b). The measurement conditions: $17kHz$, $1V_{rms}$ excitation signal.

film on a $SrTiO_3$ (100) substrate. (a) shows an initially $500 \times 500nm^2$ designed feature. A process error during the electron beam writing process led to a writing of two distinct bars of $200 \times 500nm^2$. The image area is $1 \times 1\mu m^2$. (b) shows a $1.5 \times 1.5\mu m^2$ scan of a feature with an edge length of $1\mu m$, and (c) a feature with an edge of $1.5\mu m$. The common point of all features is the very high response on the border, the up polarization for the top edges and the opposite polarization for the edge on the bottom. This asymmetry could be a result from a damage induced by the dry etching process (position of the sample with respect to the plasma). The dominant polarization is still bottom to top, however, inside the feature opposite polarized regions exist. They are preferentially directly on the border or aligned on a line parallel to the edge at a certain distance (c). A possible explanation is the existence of a-domains at the bottom of the film, which induces the switching. Figure 5.14, p.129 shows schematically the possible domain configuration in a $200nm$ thick film. The a-domain is not running through the film and hence forms interfaces at $45°$ and $135°$ with $180°$ domains. Charge neutrality at the interface requires the polarization to switch from up to down. The top corner of the a-domain is at the same time the meeting point of a $180°$ domain wall. The presumed position of the hidden $90°$ domains allows to guess that the elastic relaxation at the upper part of the feature leads to a reconfiguration of $90°$ patterns.

5.3. PAFM ON STRUCTURES OBTAINED BY THE SUBTRACTIVE ROUTE

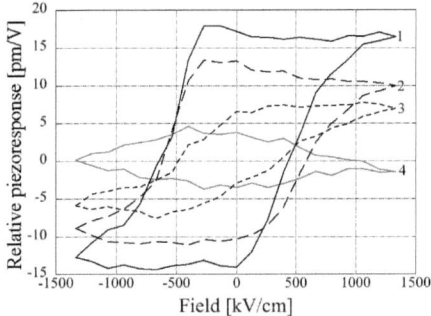

Figure 5.10: PAFM ($17kHz$, $0.4Vrms$) of $150nm$ thick features on $SrTiO_3$ (100). Measured on features with lateral size of $50nm$ (1), $100nm$ (2), $150nm$ (3), and on the unpatterned film (4).

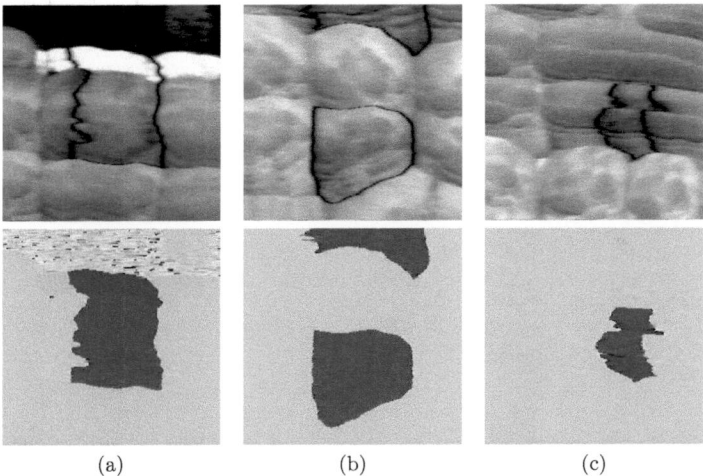

Figure 5.11: Switching experiment: PAFM images (amplitude (top) and phase (bottom)) of different features fabricated from a $200nm$ thick $Pb(Zr_{0.4},Ti_{0.6})O_3$ on $SrTiO_3$ (100). (a) $1 \times 1\mu m^2$ scan on $100 \times 300nm^2$ $Pb(Zr,Ti)O_3$ features, (b) $0.8 \times 0.8\mu m^2$ scan on triangular shaped feature, (c) $1 \times 1\mu m^2$ scan on $1000 \times 200nm^2$ long feature. The measurement conditions: $17kHz$, $1V_{rms}$ excitation signal.

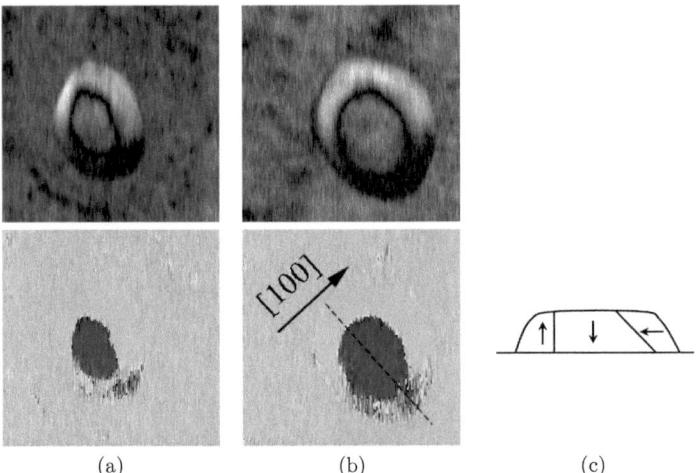

Figure 5.12: $400 \times 400 nm^2$ PAFM images of domain patterns with amplitude (top) and phase (bottom). $70nm$ (a) and $130nm$ dot from a $150nm$ thick $Pb(Zr_{0.4},Ti_{0.6})O_3$ on $SrTiO_3$ (100). The $Pb(Zr,Ti)O_3$ was only partially dry etched and $50nm$ are remaining. The measurement conditions: $17kHz$, $1V_{rms}$ excitation signal. (c) shows the proposed domain structure along the dotted line indicated in (b).

5.3. PAFM ON STRUCTURES OBTAINED BY THE SUBTRACTIVE ROUTE

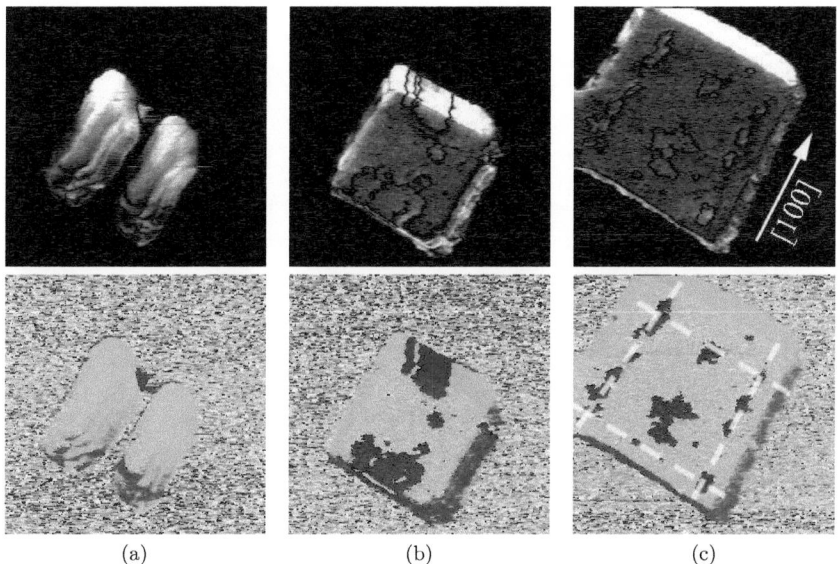

(a)   (b)   (c)

Figure 5.13: PAFM images of large scale features fabricated out of a $200nm$ thick $Pb(Zr_{0.4},Ti_{0.6})O_3$ film. (a) $1 \times 1 \mu m^2$ scan of $500nm$ features, (b) $1.5 \times 1.5 \mu m^2$ scan of feature with $1 \mu m$ edge length, (c) $2 \mu m$ scan of feature with lateral dimension of $1.5 \mu m$. The measurement conditions: $17 kHz$, $1 V_{rms}$ excitation signal.

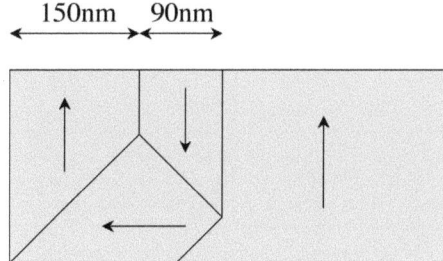

Figure 5.14: Schematic cross-section of the domain pattern observed in Figure 5.13(b-c), p.129.

5.4 PAFM on Structures Obtained by the Additive Route

This section discusses the piezoelectric response of the crystallites obtained by the additive route. First, the crystallites which nucleated without PbTiO$_3$ seed layer prior to Pb(Zr,Ti)O$_3$ deposition, and then the depositions with the application of a PbTiO$_3$ seed layer are adressed.

5.4.1 PAFM on Features Obtained without PbTiO$_3$ Starting Layer

First, the PAFM results obtained by the additive route for Pb(Zr,Ti)O$_3$ on TiO$_2$/Pt/SrTiO$_3$ (111), and then the deposition of Pb(Zr,Ti)O$_3$ on TiO$_2$/Pt/Ir/MgO (100) will be discussed.

5.4.1.1 PAFM on Features Obtained on Pt (111)/SrTiO$_3$ (111)

Figure 5.15(b, c), p.131 shows the PAFM images acquired on a large scale feature. The same features was also observed by SEM (a, top). The PAFM images show scans of $1 \times 1 \mu m^2$ (b) and $400 \times 400 nm^2$ (c). The triangles have a lateral size of typically $50 nm$ and are $15 - 20 nm$ in height. The amplitude image is shown on top and the phase on the bottom. As can be seen from the SEM image, there is space between the crystallites where the response should be zero. However, in the PAFM images this is not observed, and a measurable response is seen on the whole feature. Outside the feature though, the response is zero. The tip shape very much influences the amplitude image. The absence of zero amplitude can be attributed to the fact, that the tip doesn't reach the bottom and is always in contact with piezoelectric material. It is difficult to identify the triangles from (a) in (b). A high piezoresponse is observed at the border of individual features. The signal there is up to four times larger than the average response. From (c) it can be seen that the bright regions correspond to a down polarization while the darker contrast corresponds to the up polarization. This strong imprint is reflected in the loop measurement seen in (a, bottom), which shows a strong shift upwards. The remanent d_{33} for the up polarization is $-4.5 pm/V$ while the opposite polarization shows a d_{33} of $6 pm/V$. It is noted here, that the slight hump on the down polarization side arises from domain contributions. Compared with Figure 5.8, p.125 for the piezoresponse from the subtractive route, the aspect ratio of the triangles is at best 0.4 (for a typical lateral size of $50 nm$ and a height of $20 nm$). This would correspond to a piezoelectric response which is essentially equal to the film. However, the measured piezoelectric response is slightly higher for the naturally grown triangles. For (111) oriented features, one would expect a much lower piezoelectric response, due to the presence of a complex domain configuration as observed on $300 nm$ thick Pb(Zr$_{0.4}$,Ti$_{0.6}$)O$_3$ films (see Figure 3.31, p.74). Such a complex domain configuration is expected to lead to a pinning of domains and thus hindering the domain wall movement. However, this complex domain structure appears only on top of the $300 nm$ thick film, and was not observed on the bottom. This might be an indication, that the very thin films of $20 nm$ do not show such this domain pattern and thus show a larger piezoresponse. PAFM measurement were also done on as grown single triangles (see Figure 5.16(a), p.131 for SEM) grown on individual spots. Figure 5.16(b), p.131 shows the amplitude image with a size of $2 \times 2 \mu m^2$ and (c) the corresponding phase image. To acquire these images a tip with a much smaller tip radius was used. It is possible to see the triangular nature of the features and moreover to distinguish several triangles nucleated on a single spot. The preferred polarization is downwards, but several domains can be seen within one crystal. As before, the amplitude is not uniform and bright spots can be seen. Loop measurements on these single triangles are shown in (d). A striking feature is the appearance of a hump at the coercive

5.4. PAFM ON STRUCTURES OBTAINED BY THE ADDITIVE ROUTE

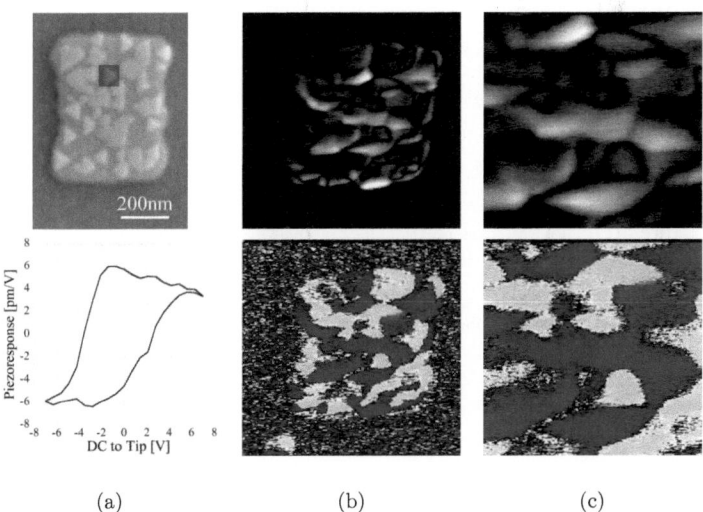

(a)　　　　　　　　(b)　　　　　　　　(c)

Figure 5.15: PAFM scans of structure obtained by the additive method without preliminary PbTiO$_3$ seed. (a) SEM image, and PAFM images (amplitude top, phase bottom), (b) $1 \times 1 \mu m^2$, (c) $400 \times 400 nm^2$ zoom into (b). The PAFM images were acquired at $1V_{rms}$ and $17kHz$ AC signal.

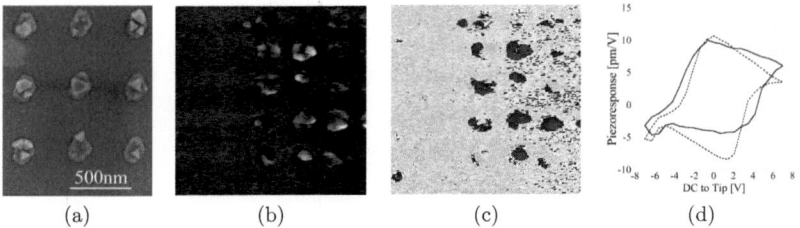

(a)　　　　(b)　　　　(c)　　　　(d)

Figure 5.16: PAFM scans on triangles obtained by the additive method without preliminary PbTiO$_3$ seed. (a) SEM image, and $2 \times 2 \mu m^2$ PAFM images: (b) amplidude, (c) phase. The PAFM images were acquired at $1V_{rms}$ and $17kHz$ AC signal. The loops in (d) were acquired at $1V_{rms}$, 17kHz.

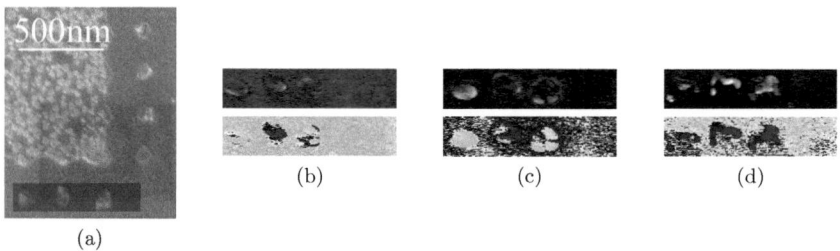

Figure 5.17: PAFM scans on triangles obtained by the additive method without preliminary PbTiO$_3$ seed. (a) SEM image, (b) - (d) PAFM images with amplitude (top) and phase (bottom). The measurement conditions: $17kHz$, $1V_{rms}$. (b) as deposited, in (c) and (d) an additional DC bias has been applied to the bottom electrode during the scanning procedure: (c) $+2V$, (d) $-6V$.

field for both polarizations. This hump is responsible for the bright spots seen in (b). Again, a slight shift upwards can be identified. The overall d_{33} is between 5 and $7pm/V$ at saturation and around $10pm/V$ at remanence, which is higher than in the case of a continuous (100) thin film on Nb-doped SrTiO$_3$ (100). Figure 5.17, p.132 shows an SEM image (a) and the corresponding PAFM images (b–c) obtained on individual triangles on TiO$_2$ dots (top: amplitude, bottom: phase image). The scanned region in (b–d) is indicated in (a). (b) shows the triangles in the as deposited state. (c) shows the piezoresponse when applying an additional DC voltage of $+2V$ to the bottom electrode, and (d) when applying $-6V$ during the scan. Therefore, switching is not expected, but an increase of the piezoresponse for already up polarized regions, and a decrease for features poled in the opposite direction. This is observed in the experiment. The left dot, which is initially polarized upwards, shows an increase in the piezoresponse. This can also be seen for the right dot. For this dot the phase is more clearly detected when applying $+2V$ to the bottom electrode. For the dot in the middle, which shows an initial down polarization , the amplitude decreases and the phase is partially undefined. Applying a voltage of $-6V$ to the bottom electrode switches the polarization. The switched down polarized regions show a higher response. But for the left dot, the switching was incomplete and the upper part remains polarized upwards. From the loop measurement in Figure 5.15(a), p.131 and Figure 5.16(d), p.131 it can be seen that switching starts at $\pm 2V$. To observe the switching behavior, scans were performed at different DC voltages from 0 to $\pm 2V$. Figure 5.19, p.134 shows a sequence of $400 \times 400nm^2$ PAFM images, acquired on some of the few randomly distributed small triangles which nucleated directly on Pt (see also SEM image in Figure 4.23(b), p.101). The topography images of the 5 triangles can be seen in Figure 5.18(a–e), p.133. The average height is $15nm$. (d) represents a sub-$100nm$ feature. In Figure 5.19, p.134, the indicated DC voltage has been applied to the bottom electrode during the scan. The following sequence was scanned: As deposited $0V$ (1), $+1V$ (2), $+2V$ (3), $+3V$ (4) back to $0V$ (5) and the same in the negative direction. The scanning speed was $0.1Hz$. In the as deposited state (a) and (c) shows a uniform polarization, the former an up polarization, the latter the opposite polarization. The other triangles show at least two opposite polarizations. Applying $+1V$ to the bottom electrode favors the bottom to top polarization. In (2) it can be seen that the amplitude of already up polarized features (a, e) is increased, and decreased for triangles having a down polarization (b–c). No switching is observed in the phase image of (d), but the up polarized regions of (e) grew and (b) clearly shows

5.4. PAFM ON STRUCTURES OBTAINED BY THE ADDITIVE ROUTE   133

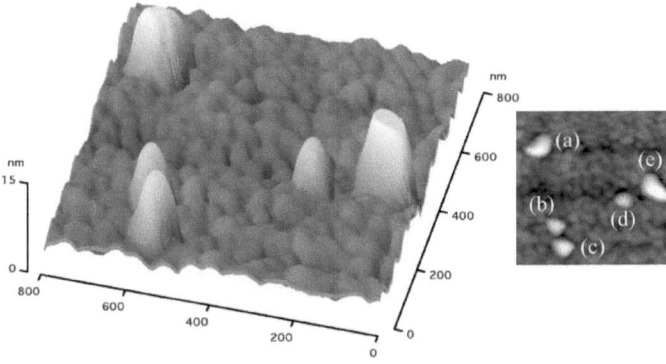

Figure 5.18: 3D AFM scan of features observed on Figure 5.19, p.134. The height of the features is about 10-15nm. The gradual switching of those features is shown in Figure 5.19, p.134.

a up polarized region with clearly defined borders in the amplitude image. Further growth of up polarized regions is observed when applying $+2V$ to the bottom electrode, and (d) switched almost completely to the up state. (c) did not switch at all. (b) shows both polarizations, and even under $+2V$, the part which is polarized down shows a response that is as large as the up polarized part. Applying $+3V$ to the bottom electrode enhances the response of the already up polarized part of feature (d). However, one also observed the vanishing of formerly up polarized regions which should show an enhanced response. This is the case on features b, d and e. This backswitching effect at high field, which was also observed on features using a PbTiO$_3$ starting layer, shows negative d_{33} values for positive field and positive values for negative d_{33} (see Figure 5.31(f), p.144). Switching back to $0V$ restores the domain configuration to the one seen at $+2V$. The overall amplitude values are generally higher than the values of the as-deposited state. Compared to the as-deposited state, feature (d) completely switched and features (b) and (e) experienced an additional but not complete switching. The next step was the switching to the opposite down state. First, a $-1V$ DC potential was applied to the bottom electrode during the scan (6). Compared to (5) all features show an increase of the down state amplitude. All features show partial switching, which is most pronounced in feature (a) where half of the area switched. Feature (d) shows a small switched region on the right side which is increased throughout increasing the DC voltage to higher values. Feature (e) shows an additional switched region on the right side. Feature (c) is now completely switched to the down state showing a well defined phase everywhere. This was not the case in (5). Increasing the DC voltage to $-2V$ (7) leads to a growth of the switched down regions, and (a) is switched completely. In (e) the partial switched down regions grew together to form one region. However, switching was not complete. Feature (d) shows a growth of the down region inside the feature. In feature (d), the up polarized phase shows a decrease in the amplitude, which is further decreased upon increasing the DC voltage to $-3V$ (8). Increasing the voltage to $-3V$ during the scan further enhances the amplitude of feature (a). A typical d_{33} value for a high piezoelectric response is $6pm/V$. Feature (c) and (d) show an increase of the polarized area. The up polarized are of feature (d) is almost zero, while the down polarized area show a higher response. Feature (b)

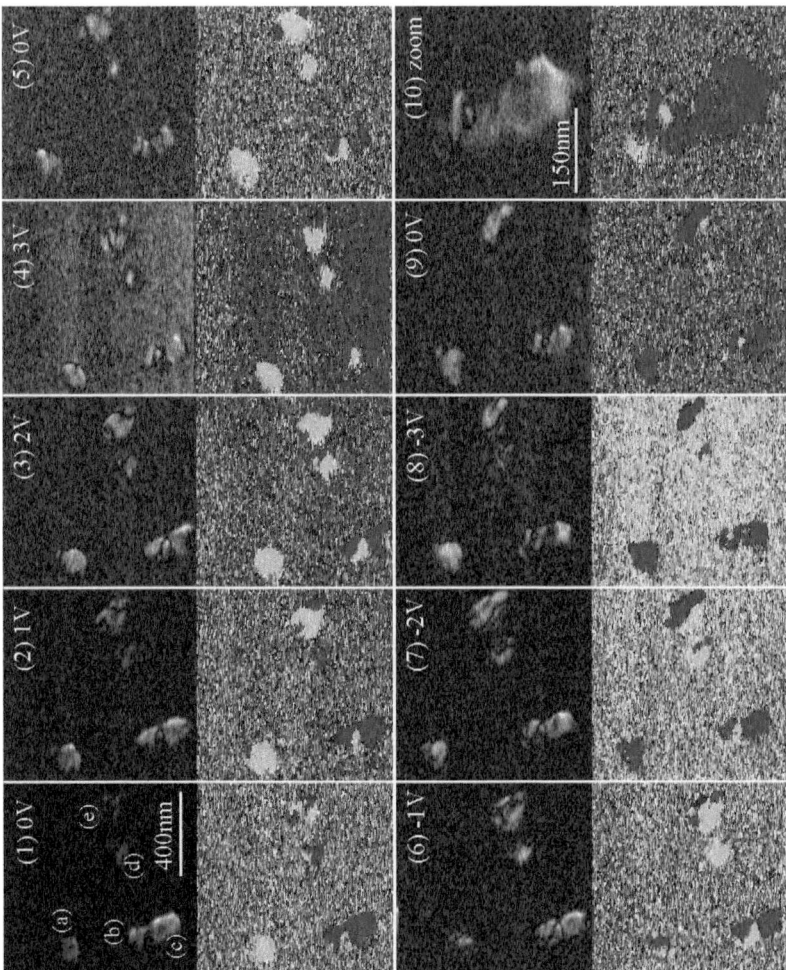

Figure 5.19: Sequence of PAFM on randomly nucleated triangles of a sample without preliminary PbTiO$_3$ deposition. Those are the same features as seen in Figure 5.18, p.133. Following the sequence, the indicated voltage was applied to the bottom electrode during the scan. The images have been acquired at $17kHz$ and $3V_{rms}$ AC signal. The sequence shows partially incomplete switching, but good retention. A backswitched domain of $15nm$ is indicated in picture (10).

5.4. PAFM ON STRUCTURES OBTAINED BY THE ADDITIVE ROUTE 135

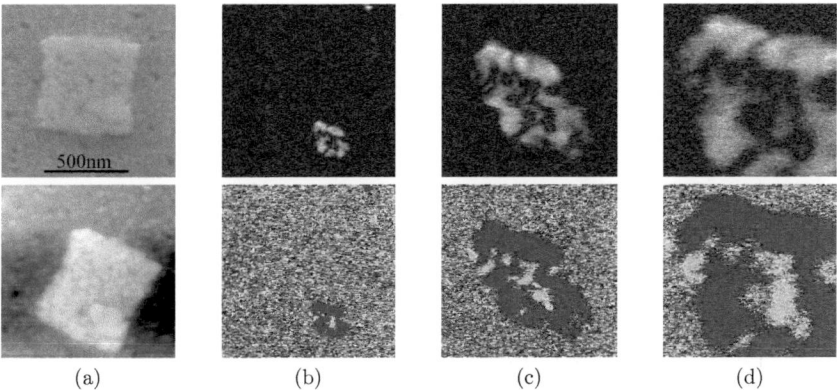

(a)　　　　　　(b)　　　　　　(c)　　　　　　(d)

Figure 5.20: PAFM scans ($2V_{rms}$ at $17kHz$) on $500 \times 500nm^2$ squared structures obtained with the additive method on $TiO_2/Pt/Ir/MgO$ (100). No $PbTiO_3$ layer was applied before $Pb(Zr,Ti)O_3$ depostition. (a, top) SEM and (a, bottom) AFM topography, (b–d) PAFM images with amplitude (top) and phase image (bottom). (b) $750 \times 750nm^2$, (c) $300 \times 300nm^2$, (d) $150 \times 150nm^2$. The brightest areas in the amplitude image correspond to a relative piezoresponse of about $4pm/V$.

shows a further increase of the down polarized area. Going back to $0V$ DC to the bottom electrode during the scan (9) leads to generally lower amplitudes. However, the amplitude of up polarized regions is decreased to zero for features (d) and (e). Only feature (c) shows remaining up polarization, where a zoom is shown in (10). The smaller domain has a diameter of $15nm$. Another observation can be made with respect to the piezoelectric response on the layer on bare Pt. This layer shows also a response, which is particularly high at high DC voltages. However, the sign of the response is inverse to the sign of the piezoelectric response, i.e. at $-3V$ DC a grey phase is observed, equivalent to a up polarization, if the layer was piezoelectric. The inverse sign can be explained by an electrostatic effect between the conductive tip and the Pt bottom electrode, where the layer on bare Pt acts as insulating capacitive interlayer.

5.4.1.2 PAFM on Features Obtained on Pt/Ir/MgO (100)

As already seen in SEM images in Figure 4.31(a), p.108, $Pb(Zr,Ti)O_3$ nucleated on some places on large scale TiO_2 seeds. PAFM images acquired on this sample are shown in Figure 5.20, p.135 and in Figure 5.21, p.136. Those Figures show, that the material is piezoelectric. The average piezoelectric response was $4pm/V$. Figure 5.20, p.135 shows a large scale feature ($500 \times 500nm^2$) with one nucleation area of $Pb(Zr,Ti)O_3$ in the bottom right corner. The SEM image (a, top) and AFM topography (a, bottom) correlate well, which indicates a sharp tip. The amplitude images are in the top row and the phase images on the bottom. (b) shows a $750 \times 750nm^2$ scan, (c) $300 \times 300nm^2$, and (d) a zoom at $150 \times 150nm^2$ scan size. A comparison between the topography (a, bottom) and the PAFM scan shows, that only the zone in the lower corner is piezoelectric. Zooming into this edge shows a feature having both, up and down polarizations. A complex domain structure can be seen, with domain sizes well below $50nm$. An explanation

136 CHAPTER 5. PIEZOELECTRIC ATOMIC FORCE MICROSCOPY (PAFM) MEASUREMENTS

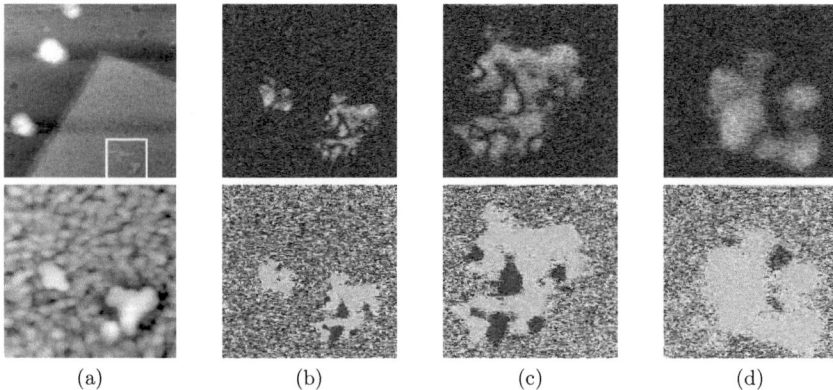

(a) (b) (c) (d)

Figure 5.21: PAFM scans on $3 \times 3 \mu m^2$ squared structures obtained with the additive method on $TiO_2/Pt/Ir/MgO$ (100). No $PbTiO_3$ layer was applied before $Pb(Zr,Ti)O_3$ depostition. AFM topographies of $2 \times 2 \mu m^2$ (a, top) and $500 \times 500 nm^2$ zoom (a, bottom) into the bottom right corner of (a, top). (b–d) PAFM images with amplitude (top) and phase image (bottom). (b) $500 \times 500 nm^2$ scan, (c) $250 \times 250 nm^2$ scan, (d) $150 \times 150 nm^2$ scan. The conditions were $2V_{rms}$ at $17 kHz$.

for this complex domain configuration might be the complex growth of $Pb(Zr_{0.4},Ti_{0.6})O_3$ (111) on $TiO_2/Pt/Ir/MgO$ (100) (see also pole figure in Figure 3.29(b), p.73). Figure 5.21, p.136 shows an SEM image (a, top), topography (a, bottom) and PAFM scans (b–d) on a very large TiO_2 seed feature. The size and location of the PAFM scan in (b) is represented in a white square in (a, top). (b) shows that only the nucleated region is piezoelectric, but not elsewhere on the large scale feature. (c) is a zoom onto the right feature, (d) on the left feature seen in (b). Again, a complex domain configuration can be seen from those images with domain sizes well below $50 nm$. PAFM scans were also performed on regions with single TiO_2 dots. Figure 5.22, p.137 shows two scans on a region where dots with a spacing of $500 nm$ have been written. (a) shows a $2 \times 2 \mu m^2$ scan, and (b) a $1 \times 1 \mu m^2$ scan, as indicated by the darker contrast square in (a). The top images represent topography images, the amplitude images are situated in the center and phase images are at the bottom. The circles in (a) represent all dots with an observable amplitude. The shape of the features is squared. Considering the complex growth of $Pb(Zr_{0.4},Ti_{0.6})O_3$ on Pt (100) with a $2 nm$ thick TiO_2 layer (see pole figure in Figure 3.29(b), p.73), it is difficult to say whether this is an epitaxial growth. However, only square shaped dots show a piezoeletric response. One edge of the squares is parallel to the MgO (100) direction as indicated in (a, top). (b) shows a zoom into (a), as indicated. Three square shaped features show a piezoelectric response. From the topography image, the [100] direction of the square's edge can be seen. The dots' edges have a length below $200 nm$. All dots show a down polarization. However, the amplitude over one dot is not uniform. This indicates again a complex domain structure within the dot. The three observed outgrows seen in top of (a, top) correspond to the bright rhombohedral spots observed with the SEM in Figure 4.30(d, type (1) rhombohedral shaped), p.107. They are not piezoelectric. However, the very large outgrow in the middle shows a small piezoresponse. Figure 5.23, p.138 shows the response of a square shaped dot in more detail. (a, top) shows the SEM image where the square shaped dot

5.4. PAFM ON STRUCTURES OBTAINED BY THE ADDITIVE ROUTE 137

is highlighted in a white square. A large scale PAFM image was taken at the same position. The corresponding topography is represented in (a, bottom). To illustrate the position of the topography (a, bottom) of the PAFM scans (b–c) with the SEM image (a, top), a rhombohedral outgrow has been highlighted with a white circle. Again, only the square shaped crystal shows a piezoelectric response. The very high outgrow in the upper part only shows a piezoelectric response on the border; inside, the amplitude is almost zero. (c) shows a $500 \times 500 nm^2$ zoom onto the square shaped crystal. The as-deposited polarization is down. Again, the amplitude is non-uniform over the feature. On the right side the amplitude is almost zero leading to an undefined phase. This could indicate a domain configuration which clamps the piezoelectric response.

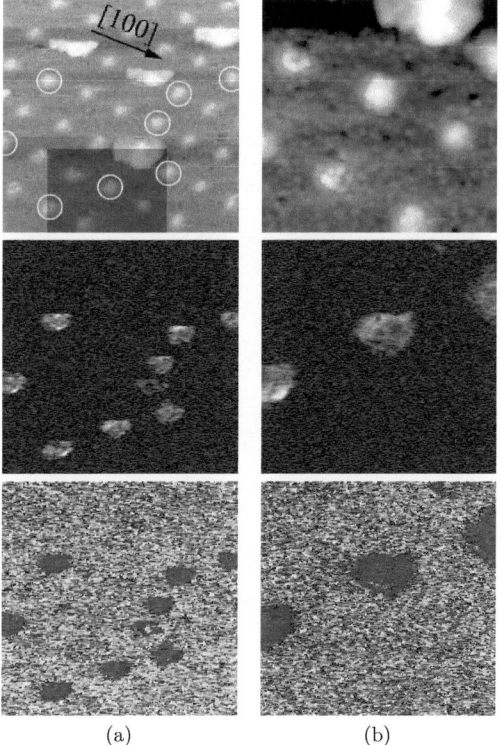

Figure 5.22: PAFM scans on single dot structures obtained with the additive method on $TiO_2/Pt/Ir/MgO$ (100). The center-to-center spacing of the dots is $500 nm$. No $PbTiO_3$ starting layer was applied before $Pb(Zr,Ti)O_3$ depostition. (a) $2 \times 2 \mu m^2$ PAFM scans, (b) $1 \times 1 \mu m^2$ PAFM scans. Top row: topography, center row: amplitude image, bottom row: phase image. The PAFM conditions were $2V_{rms}$ at $17 kHz$.

CHAPTER 5. PIEZOELECTRIC ATOMIC FORCE MICROSCOPY (PAFM) MEASUREMENTS

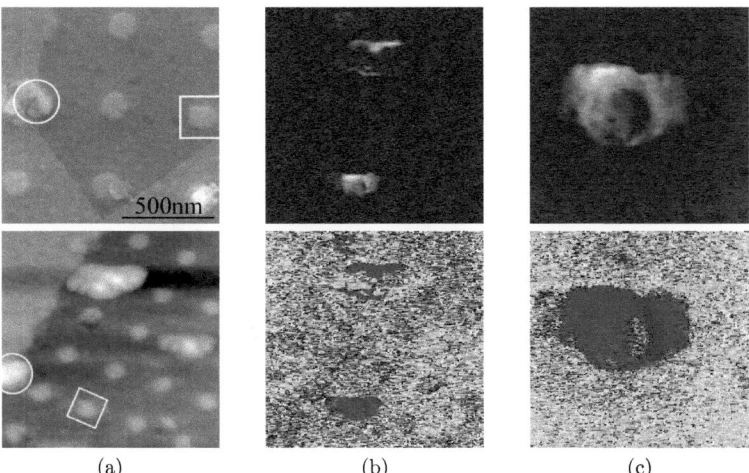

Figure 5.23: PAFM scans on structures obtained with the additive method on TiO$_2$/Pt/Ir/MgO (100). No PbTiO$_3$ layer was applied before Pb(Zr,Ti)O$_3$ deposition. SEM image (a, top) and $1.5 \times 1.5 \mu m^2$ AFM topography (a, bottom), (b) $1.5 \times 1.5 \mu m^2$ and (c) $500 \times 500 nm^2$ PAFM images with amplitude (top) and phase (bottom). The SEM image shows a cubic shaped structure, which gives a piezoelectric response in (b). The shades zone in the SEM image corresponds to the scan of the topography in (a, bottom). (b) $1.5 \times 1.5 \mu m^2$ PAFM corresponding to topography in (a, bottom), (c) zoom on the lower feature ($500 \times 500 nm^2$ scan). The conditions were $2V_{rms}$ at $17kHz$.

5.4. PAFM ON STRUCTURES OBTAINED BY THE ADDITIVE ROUTE

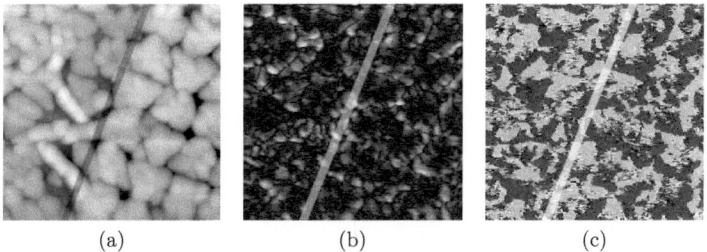

Figure 5.24: $1 \times 1\mu m^2$ PAFM scan showing Pb(Zr,Ti)O$_3$ on the bare Pt and on the patterned region. (a) topography with indicated border line between patterned (right) and unpatterned (left) region, (b) amplitude image, (c) phase image. The scan was done at $1V_{rms}$ and $17kHz$.

5.4.2 PAFM on Features Obtained with PbTiO$_3$ Starting Layers

In this route with PbTiO$_3$ seed, the samples prepared were all on SrTiO$_3$ (111) with epitaxial Pt and $2nm$ thick TiO$_2$ seed islands. First, the results of samples with a $1nm$ PbTiO$_3$ seed layer prior to the deposition of Pb(Zr,Ti)O$_3$, and then the samples with $2nm$ PbTiO$_3$ seed layer are discussed.

5.4.2.1 PAFM on Features with 1nm PbTiO$_3$ Starting Layer

In this case, the application of the PbTiO$_3$ seed led to dense nucleation as well on bare Pt and on the TiO$_2$ seeds. Figure 5.24, p.139 shows a $1 \times 1\mu m^2$ PAFM scan of the dots obtained which are spaced by $200nm$, (a) is the topography, (b) the amplitude images and (c) the phase image. There is no difference in the amplitude and phase image between the depositions on the dots (right side) and the depositions on bare Pt (left). Bright contrasts correspond to a high d_{33} of $15pm/V$. The rod-like nucleated material on bare Pt shows the same piezoelectric response. Both, up and down polarized regions can be seen. In the as-deposited state, the

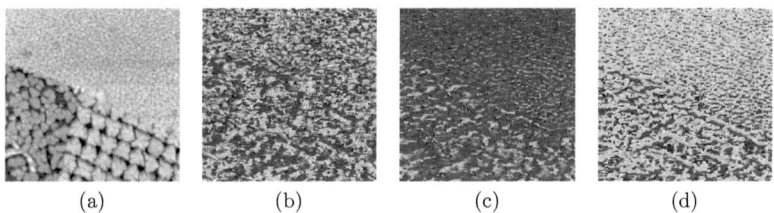

Figure 5.25: $2 \times 2\mu m^2$ PAFM topography and phase scans showing the ferroelectric switching of features obtained by adding a starting layer of $1nm$ PbTiO$_3$. (a) topography (top and bottom right), and bare Pt (bottom left), (b) phase image as deposited, (c) phase image obtained at $+4V$ to the tip during scan, and (d) phase image obtained at $-4V$. The tip signal was $1V_{rms}$, $17kHz$.

140 CHAPTER 5. PIEZOELECTRIC ATOMIC FORCE MICROSCOPY (PAFM) MEASUREMENTS

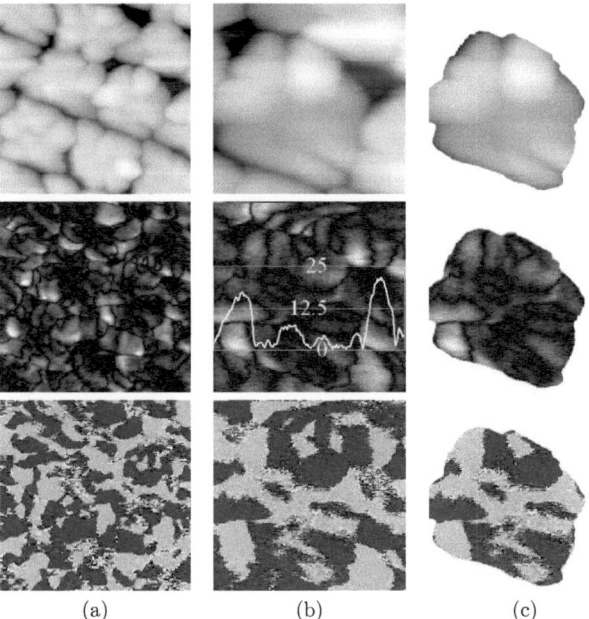

(a) (b) (c)

Figure 5.26: PAFM scans of features obtained by the additive method with a $1nm$ thick PbTiO$_3$ starting layer. Topography (top), amplitude (middle) and phase contrast (bottom). (a) $500 \times 500 nm^2$ scan, (b) $250 \times 250 nm^2$ scan, (c) enhanced feature. The amplitude image in (b) shows a line scan of the relative piezoresponse. The condition was $1V_{rms}$ and $17 kHz$.

frequency of both polarizations is the same. Figure 5.26, p.140 shows zooms onto Pb(Zr,Ti)O$_3$ which nucleated on the dots. (a) shows a $500 \times 500 nm^2$ scan, (b) a $250 \times 250 nm^2$ scan, and in (c) one dot from (b) has been cut to illustrate the domain structure within one nucleated feature. The image on top is the topography, in the middle the amplitude, and below is the corresponding phase image. In (b, middle), a line scan showing the piezoelectric response over a line scan. A high piezoelectric response between 5 and $22 pm/V$ was observed. From the topographic image in (c) the individual grains of which the whole feature is composed of can be seen. Each grain is composed of several domains. The smallest domain has a lateral size of less than $50nm$. Switching could be obtained by applying $\pm 4V$ to the bottom electrode during the scan. This is shown in Figure 5.25, p.139. (a) shows the topography showing the regions where Pb(Zr,Ti)O$_3$ nucleated onto bare Pt, onto the TiO$_2$ dots and onto a large scale TiO$_2$ area. (b–d) show the phase images, where (b) is the as-deposited state, (c) when applying $-4V$ to the bottom electrode, and (d) when applying $+4V$ to the bottom electrode. Comparing the contrasts in (b–d), it can be seen that switching took place. However, some regions remained always in the opposite direction. This is because the applied voltage was not high enough for complete switching.

5.4. PAFM ON STRUCTURES OBTAINED BY THE ADDITIVE ROUTE

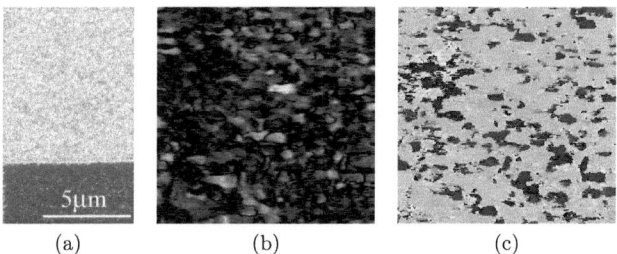

(a) (b) (c)

Figure 5.27: PAFM scans of PZT features obtained by the additive method with a $2nm$ thick PbTiO$_3$ starting layer on $2nm$ TiO$_2$ (patterned)/Pt/SrTiO$_3$ (111). $3 \times 3\mu m^2$ PAFM scans on a region where a large area of TiO$_2$ existed. (a) SEM image, (b) amplitude, (c) phase (gray = polarization up, black = polarization down). The PAFM images have been acquired at $1V_{rms}$ and $17kHz$ AC signal. The piezoelectric response varies between 2 and $10pm/V$. The average value is $2.5pm/V$.

5.4.2.2 PAFM on Features Obtained with 2nm PbTiO$_3$ Starting Layer

In this process, a $2nm$ PbTiO$_3$ seed was applied prior to Pb(Zr,Ti)O$_3$ deposition. Figure 5.27, p.141 shows an SEM image of a large area TiO$_2$ surface covered with Pb(Zr,Ti)O$_3$ (a). (b) shows the PAFM amplitude image and (c) the phase. The piezoelectric amplitude ranges from 3 to up to $5pm/V$. The average d_{33} value is $2.5pm/V$. The majority of the crystals shows an polarization. The overall piezoelectric response is generally lower than on all the other samples. PAFM measurements on single crystal features are shown in Figure 5.28, p.142. (a, top) shows the topography image. The inset represents a line scan over the features. Their height is between 10 and $30nm$. (a, bottom) shows a ferroelectric loop acquired in the center of feature shown in (c). A shift upwards, *i.e.* towards down polarization can be seen. The remanent d_{33} values are $-2pm/V$ and $3.5pm/V$. The PAFM scans on the as grown features often show an inverse phase in the center (see Figure 5.28(b–c), p.142, Figure 5.29(a,c), p.143 and Figure 5.30(a–d), p.143). The piezoelectric response of the features is down on the border and up in the center. The response in the down polarized center is lower than on the border. In the center, d_{33} of typically $3.5pm/V$, and at the border, values between $3-5pm/V$ are measured. In the phase image, the transition from the up polarization in the center to the down polarization at the border is smeared out and happens over several tens of nanometers. This is in contrast to 180° domain walls where abrupt changes of the phase are observed. The smeared out transition can be explained by the existence of *a*-domains in the center or in the transition region. An *a*-domain itself does not contribute to the d_{33}, and moreover, *a*-domains may hinder *c*-domains from expanding through clamping. The strong response of the white spot in (a) and (b) arises from a triangle that nucleated on bare Pt. Its height is $10nm$. As can be seen from the $150 \times 150nm^2$ scan in (b) the size of the triangle must be well below $100nm$ taken into account the size of the tip. The response amounts to $5.5pm/V$. An overview of the different domain patterns in square shaped crystals is given in Figure 5.30(a–d), p.143. (a) shows a $200 \times 200nm^2$ scan, and (b–d) are $300 \times 300nm^2$ scans. Typically, less than 10 domains can be observed within one crystal. Sharp domain walls can be seen in (c) and (d) indicating 180° domains. On both sides of such sharp walls, the piezoresponse is generally higher than for the smeared out transition. (d) shows a high piezoresponse not only at the border, but also inside the crystal. The smallest domain width is below $20nm$. As already observed in Figure 5.27, p.141 very bright spots were also found on the crystals (Figure 5.30(c−d), p.143). Figure 5.31, p.144 shows three such "hot spots" observed on TiO_2 seeds which were $500 \times 500nm^2$ large. Three spots are indicated. The amplitude of the other regions is typically $2.5pm/V$, and on the bright spots, an amplitude of up to $10pm/V$ is deduced from the amplitude image (b). In the as-grown state, the spots can be both , up and down polarized (c). PAFM loop measurements performed on those spots revealed very high d_{33} are shown in the images below (1−3). Values of up to $10pm/V$ were found at zero field. This corresponds to the values found in the amplitude images. All measurements have the hump around the remanent polarization in common. This hump can be explained by some domain effects. Another common point is the linear decrease of the piezoresponse from the hump to the saturation voltage. In (2), the hump accounts for a double of the piezoelectric response at zero external field. A curiosity was observed on point (1) and (3). The loops measured at an excitation signal of $0.5V_{rms}$ showed an inverse polarization at saturation, i.e. the d_{33} value was positive for negative voltages and vice versa. Switching took place when reducing the voltage towards the opposite sign. Increasing the excitation voltage drove the system towards "normal" loops. This can be seen from the measurements on spot (3). When applying $1.5V_{rms}$, the d_{33} at saturation was just zero, and increasing further to $2V_{rms}$ led to a "normal" loop. Figure 5.32, p.145 shows a selection of acquired loops on $500 \times 500nm^2$ seed islands. Even though $Pb(Zr,Ti)O_3$ grows epitaxially in the (111) direction on TiO_2/Pt (100), a large variety of loop-shapes exist indicating complex domain configurations. The highest d_{33} values are up to $10pm/V$. Loops (2) and (3) represent typical loops on this sample showing a d_{33} of up to $3pm/V$ at saturation and at remanence. It is worth noting here, that the loop (2) and (3) were acquired on features which were only 6 and $8nm$ high.

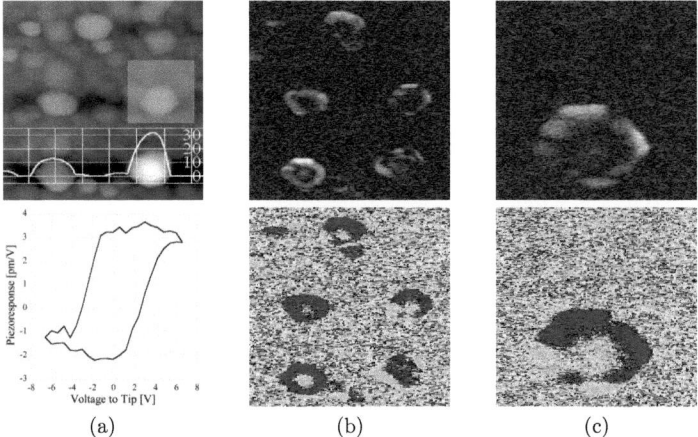

Figure 5.28: PAFM scans of PZT features obtained by the additive method with a $2nm$ thick $PbTiO_3$ starting layer on $2nm$ TiO_2 (patterned)/Pt/$SrTiO_3$ (111). PAFM scans on square shaped structures obtained with the additive method (SEM see Figure 4.29, p.105). (a) topography with indicated feature hight. The inset shows the region, where (c) was taken. (b) $1 \times 1\mu m^2$ and (c) $750 \times 750nm^2$ PAFM scans (amplitude up, phase down) of squared features obtained at $17kHz$, $2V_{rms}$.

5.4. PAFM ON STRUCTURES OBTAINED BY THE ADDITIVE ROUTE 143

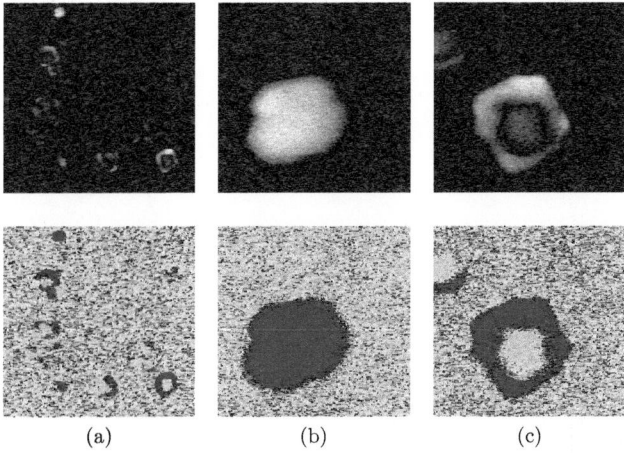

Figure 5.29: PAFM scans on features obtained by the additive method with a $2nm$ PbTiO$_3$ starting layer (SEM see Figure 4.29, p.105). Top: amplitude image, bottom: phase image. (a) $1.2 \times 1.2 \mu m^2$, (b) $150 \times 150 nm^2$ detail of bright spot showing a d_{33} of $12.5 pm/V$. (c) $300 \times 300 nm^2$ scan of feature down right in (a). The border shows a d_{33} of $6-10 pm/V$, the inner part an average of $3.5 pm/V$. The PAFM images were acquired at $2V_{rms}$ and $17kHz$ AC signal.

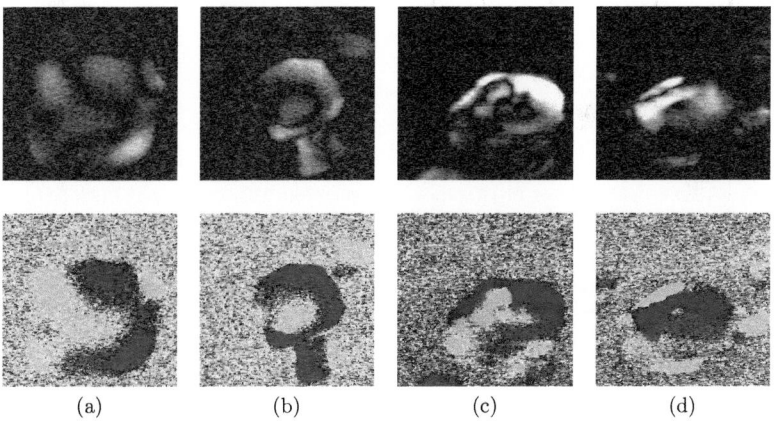

Figure 5.30: PAFM scans on features obtained by the additive method with a $2nm$ PbTiO$_3$ starting layer (SEM see Figure 4.29, p.105). (a) $200 \times 200 nm^2$, (b–d) $300 \times 300 nm^2$. The PAFM images were acquired at $2V_{rms}$ and $17kHz$ AC signal.

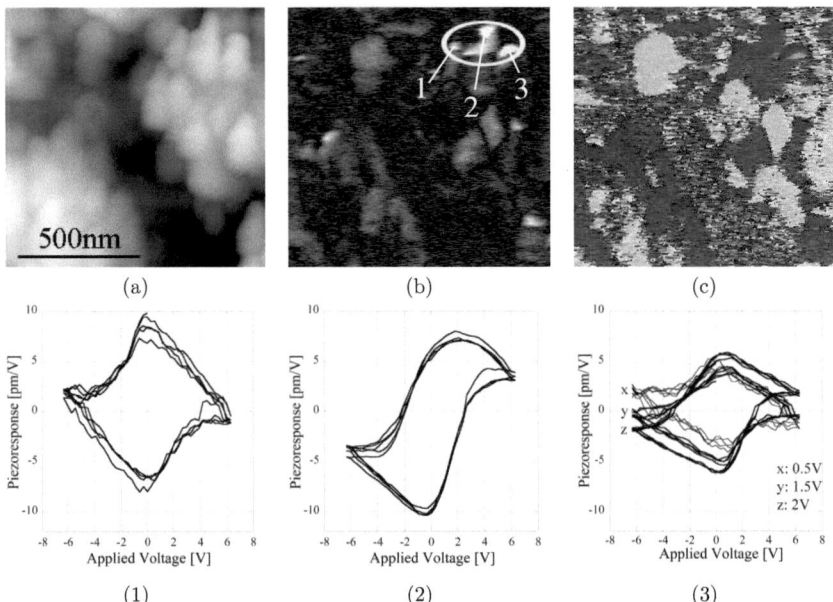

Figure 5.31: PAFM loop measurements on PZT features obtained by the additive method with a $2nm$ thick PbTiO$_3$ starting layer on $2nm$ TiO$_2$ (patterned)/Pt/SrTiO$_3$ (111). $1 \times 1 \mu m^2$ PAFM scans showing Pb(Zr,Ti)O$_3$ on $500 \times 500 nm^2$ seeds (see Figure 4.27, p.104 for SEM image). (a) topography, (b) amplitude and (c) phase image. The white circle in (b) shows an area with 3 locations of extremely high piezoresponse. Local loops: (1) at position 1 acquired at $0.5V_{rms}$, (2) $2V_{rms}$ at position 2, (3) at position 3 for different voltages (x, y, z indicated in rms). The frequency was $17kHz$.

5.4. PAFM ON STRUCTURES OBTAINED BY THE ADDITIVE ROUTE

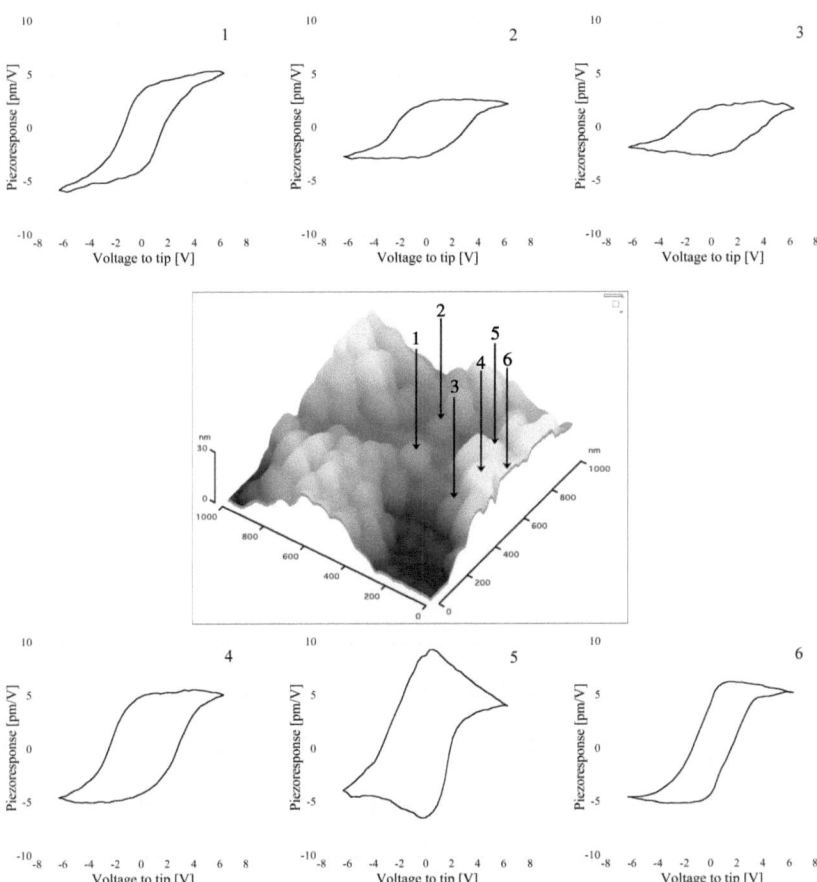

Figure 5.32: PAFM loop measurements on PZT features obtained by the additive method with a $2nm$ thick $PbTiO_3$ starting layer on $2nm$ TiO_2 (patterned)/Pt/$SrTiO_3$ (111). Local PAFM loops on grains of $500 \times 500 nm^2$ seeds. The height of the particles are 1: $8nm$, 2: $6nm$, 3: $8nm$, 4: $22nm$, 5: $25nm$, 6: $37nm$.

5.5 Summary and Conclusion

The small features fabricated by subtractive and additive routes were characterized by piezoelectric AFM measurements. The smallest obtained features were still ferroelectric, and piezoelectricity could be measured on features as thin as $6nm$. The main difference between the features obtained by the subtractive and the additive method is the much higher aspect ratio in the former case. In the additive route, the aspect ratio is much lower, and a quasi-continuous thin film situation can be supposed. In the case of continuous $100nm$ thick films, the piezoelectric response was typically $5pm/V$, which is in the typical range of piezoelectric response measured on $20nm$ thick features obtained by the additive route on (111) substrates. As the aspect ratio of the feature is increase to above 1, an increased piezoelectric response was observed. In addition to un-clamping, we propose a change in the domain configuration as the driving force behind this increase, $i.e.$ depinning of domains and/or vanishing of a domains. This hypothesis is supported by training experiments on continuous $200nm$ thick films. After applying alternating DC pulses well above the coercive voltage, the local piezoelectric response on the film is temporarily up to 5 times larger than without the training. The piezoresponse after training is comparable to the response observed for patterned high aspect ratio features. Theory shows that the increase in d_{33} is related to a dramatic change in the domain configuration and that it corresponds well with the sum of contributions from the clamped film ($105pm/V$) plus the contribution of mobile a-domains ($60pm/V$). Border effects were observed on (100) features obtained by both routes. In both cases, an opposite polarization was observed in the center of the features. The domain pattern could be explained by the presence of a domains near the border which induce the down polarization in the center. The size of the induced downwards polarized domains is depending on the features size. For $200nm$ thick features obtained by the subtractive route, the size of the down polarized domain was typically $150nm$ whereas for $20nm$ thick features obtained by the additive route, the size was below $50nm$. Domain contributions to the piezoelectric response were important as well for measurements performed on features with a high aspect ratio and on features obtained by the additive route. In both cases, PAFM d_{33} measurements showed a hump in the ferroelectric loop around zero field due to 180° domain switching leading to a large piezoelectric response. For features obtained by the additive route, this was locally observed on some points on triangular and square shaped crystals with d_{33} values up to $10pm/V$. A large variety of loops was acquired on different local positions on features obtained by the additive route. Different loop shapes were obtained pointing out different domain configurations. $20nm$ thick features obtained on TiO_2/Pt (111) showed complex domain patterns. Up to 10 domains were found in features with a lateral size of $100nm$. Sharp domain walls were identified (180° domain wall) and transitions showing a gradual change in the amplitude which passes through zero. The latter case was identified to be due to buried a domains which run at 45° towards the top of the feature. The smallest domains observed were 180° domains and had a width of $15nm$.

General Conclusions

The goal of this work was to study suitable processes to achieve damage-free ferroelectric features of less than $100nm$ diameter whose positions are defined by a lithography step, and to characterize their ferroelectric properties. This study was motivated by needs for downscaling in microlectronics on the one hand, and open questions on critical dimensions related to domain patterns and the existence of ferroelectricity on the other hand. The investigated processes were based on in-situ sputter deposited $PbZrTiO_3$ thin films, a material that looks most promising for memory applications. In order to obtain a good definition of growth substrates, the study was carried out using epitaxial growth on $SrTiO_3$ crystals.

Damage-free ferroelectric structures have been obtained by means of two different routes. In a first route, relatively thick PZT films have been patterned using e-beam lithography with a positive resist (PMMA) to structure a Cr masking layer. The Cr pattern was transferred to PZT by a dry etching process using a ECR-*RF* dual frequency tool exhibiting low ion energies and high ion fluxes at low pressures. The smallest fabricated features were $80nm$ wide, and $200nm$ high. Limiting factors were the backscattering of secondary electrons from the PZT film, and difficulties encountered with the lift-off structuration of the Cr film. In a second route the PZT features were obtained using lithography-modulated self-assembly of PZT. A $2nm$ thick affinity layer of $TiO_2(100)$ grown on $Pt(111)/SrTiO_3$ (111) was patterned into dots by e-beam lithography using preferentially a negative e-beam resist (Calixarene), which serves directly as etching mask. The resolution in this case was limited by backscattering of secondary electrons generated in the Pt bottom electrode. The smallest TiO_2 features amounted again to about $70nm$. PZT preferentially nucleated on the TiO_2 islands. Two growth modes have been identified. If the process starts lead-rich and without Zr flux dring the first 2 nm, the obtained PZT feature covers the affinity spot and grows as quasi square plates in (100) orientation, indicating a layer-by-layer growth. If, however, the process starts with less lead excess and including Zr, triangular PZT(111) crystallites are observed that do not cover the complete affinity spot, indicating that island growth is dominating. The effect of the affinity spots could be well shown. Nucleation on the spots is very likely, and the near surrounding of the spots does not contain other nuclei. The diffusion phenomena could be well explained with a complete rate equation for PbO adatoms that includes the rates of deposition, desorption, chemisorption, and diffusion. The (100) growth mode is a result of a large PbO flux onto the seed leading to nucleation under PbO rich condition. This is situation is a consequence of a large desorption rate on the affinity spot that exceeds the desorption rate on the bare platinum. The presence of Zr inverts the situation. Desorption on platinum becomes larger than the chemisorption on the affinity spot. As a result, the latter looses lead to the platinum resulting in lead poor conditions for nucleation, and a switching to (111)-orientation promoted by the TiO_2 (100) affinity layer. The experiments clearly show that the lead excess of the process needs to be adjusted for a nucleation probability equal to 1 for each affinity spot.

A number of interesting phenomena related to domains have been observed by means of piezo-sensitive AFM. Extended PZT films show strong self-poling leading to a homogeneous phase and amplitude contrast. X-ray diffraction as well as TEM images reveal the existence of minority 90° domains, with typical widths of $10nm$. However, that hardly can be detected by PAFM. Patterned films (subtractive method) with relatively large aspect ratios (height/width) show an increased response at borders, which extends to the whole feature if the aspect ratio is larger than one. One might argue that the clamping effect is reduced and the response thus increased. However, the size of the signal increase, the piezoelectric loop, as well as the observed domain patterns indicate that 90° domain walls are removed from the upper edges, leading to an additional signal increase. The 90° domains are not needed anymore to reduce the elastic energy in the upper part of the strained clamped film, and are thus removed during the etching process, meaning that they were not fixed by defects, but by the clamping to the substrate. There is also evidence that domain walls become more mobile and contribute to the piezoelectric response. The liberation of domain walls has also been achieved by nano-training at extended films, meaning the ferroelectric switching induced by voltage cycling applied to the AFM tip. The signal increase was as large as obtained with large aspect ratios but decayed quickly after stopping the cycling. In this case the clamping to the substrate together with constraints imposed by domains outside the trained zone install a domain configuration equivalent to the previous one. The signal increase observed at the high aspect ratio structures - a kind of the "nano-1-3-composite" - is thus less due to simple reduction in piezoelectric clamping, but has more to do with the release of ferroelastic domains.

The different features obtained by the various processing routes allowed also to attack the question of smallest observed domains and thinnest films exhibiting piezoelectric loops. The smallest features investigated by PAFM had diameters of around $50nm$. They still exhibited domains. All data indicate so far that the smallest domains had diameters of around $20nm$. They were observed at $20nm$ thick films. We could not establish a good statistics to have evidence that the thickness dictates the size of the smallest domain. However, it is clear that a relation of this kind must exist. The PAFM loop amplitude clearly decreased with film thickness. The thinnest feature still showing a clear ferroelectric switching was $6nm$ thick and was grown on Pt (111). An attempt was made to compare experimental findings with the Landau-Devonshire-Ginzburg phenomenological theory. The $d_{33,f}$ loop together with the CV-loop of a $600nm$ thick, mostly c-oriented film could be fairly well be described by the theory. Deviations could be identified as domain wall contributions. The critical thickness for ferroelectricity was estimated using in addition literature data for domain wall energies. The critical thickness was calculated as 1 to $2nm$, which does not contradict experimental findings.

Outlook

There are two important questions arising in the fabrication of future ferroelectric memories:

- How far will downsizing be possible ?
- How will the etching process influence ferroelectricity at these sizes ?

This work gave answers to these questions. First of all, the ultimate limit of ferroelectricity could not be reached at the sizes obtained. On the one hand this is regrettable from the research point of view, but on the other hand this is good news for the fabrication of memories. Ferroelectricity was clearly detected in thin-film-like flat features as thin as $6nm$. These features still showed complex domain patterns, indicating that they were far away from a single domain state. Moreover, the PAFM piezoresponse of $20nm$ thick (111) features was surprisingly high and comparable to the one of $100nm$ thick PZT (001) films. From our theoretical predictions, $2nm$ thick PZT (001) films should still be ferroelectric.

The question of patterning issues also arises. We could show that the appropriate dry-etching process does not alter the ferroelectric properties and damage-free sub-$100nm$ capacitors have been obtained. Furthermore, we showed, that patterning is even favorable to the ferroelectric properties: High aspect ratio capacitors showed enhanced piezoelectric properties. This is probably due to a change in the domain structure from a to c-domain when the aspect ratio starts to become larger than 1. It is expected, that further downscaling amplifies this effect. This can be regarded as a step towards single domain capacitors. Such an effect counteracts the scattering of properties in very small capacitors made of only few grains. This is not only interesting for epitaxial ferroelectric films, but it could also be exploited in oriented polycrystalline thin films. In the additive route, we demonstrated an alternative way for the fabrication of defect-free small ferroelectric capacitors, which does not need the patterning of the ferroelectric layer. The good functioning of this process opens the possibility to achieve a kind of self-aligned PZT deposition for high-density memories, allowing a reduction of the number of lithography steps. In contrast to the subtractive route, this process produces flat features. This is particularly interesting for the fabrication of patterned ferroelectric media. If probed by an AFM tip, the advantage of such a device is the easy detection of a bit's position, which is given by the topography.

Finally, this additive route is a promising technique which is not restricted to ferroelectrics, It simplifies the processing and generates defect free functional features. The conditions are a long residence time of the most volatile species on the main part of the surface and its immediate capture on a seed. Possible candidates are low melting point materials like Ga, In, Sn and Bi.

Bibliography

[1] S. Bühlmann, J. Baborovski, B. Dwir, and P. Muralt. Size-effect in mesoscopic epitaxial ferroelectric structures: Increase of piezoelectric response with decreasing feature-size. *Appl. Phys. Lett.*, 80(17):3195, 2002.

[2] S. Bühlmann, P. Muralt, and S. Von Allmen. Lithography-modulated self-assembly of small ferroelectric Pb(Zr,Ti)O_3 single crystals. *Appl. Phys. Lett.*, 84(14), 2004.

[3] S. Bühlmann, P. Muralt, P. A. Kuenzi, and U. Staufer. Electron beam lithography with negative Calixarene resists on dense materials: Taking advantage of proximity effects to increase pattern density. *J. Vac. Sci. Technol. B*, 23(5):1895, 2005.

[4] S. Bühlmann, E. Colla, and P. Muralt. Polarization reversal due to charge injection in ferroelectric films. *Phys. Rev. B*, 72(21):214120, 2005.

[5] S. Bühlmann and P. Muralt. Electrical Nanoscale Training of Piezoelectric Response Leading to Theoretically Predicted Ferroelastic Domain Contributions in PZT Thin Films. *Advanced Materials*, 20(16):3090, 2008.

[6] S. Y. Wu, J. Takei, and M. H. Francombe. Optical switching characteristics of epitaxial bismuth titanate films for matrix addressed displays. *Ferroelectrics*, 10:209, 1976.

[7] H. Park, J. Jung, S. Kim, S. Hong, and H. Shin. Scanning resistive probe microscopy: Imaging ferroelectric domains. *Appl. Phys. Lett.*, 84(10):1734, 2004.

[8] Y. Cho, K. Fujimoto, Y. Hiranaga, Y. Wagatsuma, A. Onoe, K. Terabe, and K. Kitamura. Terabit inch^{-2} ferroelectric data storage using scanning nonlinear dielectric microscopy nanodomain engineering system. *Nanotechnology*, 14:637, 2003.

[9] C. H. Seager, D. C. McIntyre, W. L. Warren, and B. A. Tuttle. Charge trapping and device behavior in ferroelectric memories. *Appl. Phys. Lett.*, 68(19):2660, 1996.

[10] R. Moazzami, P.D. Maniar, R. E. Jones, and C. J. Mogab. Integration of ferroelectric capacitor technology with CMOS. *Symp. VLSI Techno. Dig. Techn. Papers*, page 55, 1994.

[11] U. Böttger and S. R. Summerfelt. *Ferroelectric random access memories*, volume Advanced electronic materials and novel devices of *Nanoelectronics and Information Technology*. Wiley-VCH, Aachen, 2003.

[12] K. Amanuma, S. Kobayashi, T. Tatsumi, Y. Maejima, H. Hada, J. Yamada, T. Miwa, H. Koike, H. Toyoshima, and T. Kunio. Characteristics of $0.25\mu m$ ferroelectric nonvolatile

memory with a Pb(Zr,Ti)O$_3$ capacitor on a metal/via-stacking plug. *Jpn. J. Appl. Phys.*, 39(Part 1, 4B):2098, 1999.

[13] S. R. Summerfelt, T. S. Moise, G. Xing, L. Colombo, T. Sakoda, S. R. Gilbert, A. L. S. Loke, S. Ma, L. A. Wills, R. Kavari, T. Hsu, J. Amano, S. T. Johnson, D. J. Vestcyk, M. W. Russell, S. M. Bilodeau, and P. van Buskirk. Demonstration of scaled (>=0.12μm) Pb(Zr,Ti)O$_3$ capacitors on W plugs with Al interconnect. *Appl. Phys. Lett.*, 79(24):4004, 2001.

[14] O. Auciello, J. F. Scott, and R. Ramesh. The Physics of Ferroelectric Memories. *Phys. Today*, 51:22, 1998.

[15] J. Rickes, A. Bartic, D. Wouters, and R. Waser. Comparison between standard and chain-type FRAM architectures. *Integrated Ferroelectrics*, 34:1, 2001.

[16] M. H. Francombe and S. V. Krishnaswamy. Growth and properties of piezoelectric and ferroelectric films. *J. Vac. Sci. Technol. A*, 8(3):1382, 1990.

[17] J. F. Scott and B. Pouligny. Raman spectroscopy of submicron KNO$_3$ films. Fatigue and space-charge effects. *J. Appl. Phys.*, 64(3):1547, 1988.

[18] K. Dimmler, M. Parris, D. Butler, and S. Eaton. Switching kinetics in KNO$_3$ ferroelectric thin-film memories. *J. Appl. Phys.*, 61(12):5467, 1987.

[19] J. F. Scott and C. A. P. de Araujo. Ferroelectric Memories. *Science*, 246:1400, 1989.

[20] M. Grossmann, O. Lohse, D. Bolten, U. Boettger, R. Waser, S. Tiedke, T. Schmitz, U. Kall, M. Kastner, G. Schindler, and W. Hartner. Influence of the measurement parameters on the reliability of ferroelctric thin films. *Integrated Ferroelectrics*, 32:1, 2001.

[21] H. Uchida, N. Soyama, K. Kaegeyama, K. Ogi, M. C. Scott, and C. A. P. de Araujo. Characterization of self-patterned SrBi$_2$Ta$_2$O$_9$ thin films from photo-sensitive solutions. *Integrated Ferroelectrics*, 16:41, 1997.

[22] K. Amanuma and T. Kunio. Low-Voltage Switching Characteristics of SrBi$_2$Ta$_2$O$_9$ Capacitors. *Jpn. J. Appl. Phys.*, 35(Part 1, 9B):5229, 1996.

[23] H. Uchida, N. Soyama, K. Kageyama, K. Ogi, M. C. Scott, J. D. Cuchiaro, L. D. McMillan, and C. A. P de Araujo. Characterization of self-patterned SBT/SBNT thin films from photo-sensitive solutions. *Integrated Ferroelectrics*, 18:249, 1997.

[24] C. Isobe, T. Ami, K. Hironaka, K. Watanbe, K. Sgiyama, N. Nagel, K. Katori, and Y. Ikeda. Characteristics of ferroelectric SrBi$_2$Ta$_2$O$_9$ thin films grown by "FLASH" MOCVD. *Integrated Ferroelectrics*, 12:95, 1997.

[25] Y. G. Wang, W. L. Zhong, and P. L. Zhang. Lateral size effects on cells in ferroelectric films. *Phys. Rev. B*, 51(23):17235, 1995.

[26] G. Arlt, D. Hennings, and G. de With. Dielectric properties of fine-grained barium titanate ceramics. *J. Appl. Phys.*, 58(4):1619, 1985.

[27] S. B. Ren, C. J. Lu, J. S. Liu, H. M. Shen, and Y. N. Wang. Size-related ferroelectric-domain-structure transition in a polycrystalline PbTiO$_3$ thin films. *Phys. Rev. B*, 54(20):14337, 1996.

[28] C. W. Chung, J. K. Lee, C. J. Kim, and I. Chung. Fabrication and comparison of ferroelectric capacitor structures for memory applications. *Integrated Ferroelectrics*, 16:139, 1997.

[29] M. Alexe, C. Harnagea, A. Visinoiu, A. Pignolet, D. Hesse, and U. Gösele. Patterning and switching of nano-size ferroelectric memory cells. *Scripta Mater.*, 44(8-9):1175, 2001.

[30] J. F. Scott, M. Alexe, N. D. Zakharov, A. Pignolet, C. Curran, and D. Hesse. Nano-phase SBT family ferroelectric memories. *Integrated Ferroelectrics*, 21:1, 1998.

[31] R. E. Jones. Integration of ferroelectric nonvolatile memories. *Sol. St. Technol.*, 40:201, 1997.

[32] H. M. Duiker, P. D. Beale, J. F. Scott, C. A. P. de Araujo, B. M. Melnick, J. D. Cuchiaro, and L. D. McMillan. Fatigue and switching in ferroelectric memories: Theory and experiment. *J. Appl. Phys.*, 68(11):5783, 1990.

[33] K. Amanuma, T. Hase, and Y. Miyasaka. Fatigue Characteristics of Sol-Gel Derived Pb(Zr,Ti)O$_3$ Thin Films. *Jpn. J. Appl. Phys.*, 33(Part 1, 9B):5211, 1994.

[34] M. Grossmann, O. Lohse, D. Bolten, R. Waser, W. Hartner, G. Schindler, C. Dehm, N. Nagel, V. Joshi, N. Solayappan, and G. Derberwick. Imprint in ferroelectric SrBi$_2$Ta$_2$O$_9$ capacitors for non-volatile memory applications. *Integrated Ferroelectrics*, 22:95, 1998.

[35] A. Tagantsev and I. Stolichnov. Injection-controlled size effect on switching of ferroelectric thin films. *Appl. Phys. Lett.*, 74(9):1326, 1999.

[36] A. K. Tagantsev. Size effects in polarization switching in ferroelectric thin films. *Integrated Ferroelectrics*, 16:237, 1997.

[37] C. A. P. de Araujo, J. D. Cuchiaro, L. D. McMillan, M. C. Scott, and J. F. Scott. Fatigue-free ferroelectric capacitors with platinum electrodes. *Nature*, 374:627, 1995.

[38] Y. Watanabe and A. Masuda. Ferroelectric Self-Field Effect: Implications for Size effect and memory device. *Integrated Ferroelectrics*, 27:51, 1999.

[39] A. Gruverman, B. J. Rodriguez, A. I. Kingon, R. J. Nemanich, J. S. Cross, and M. Tsukada. Spacial inhomogeneity of imprint and switching behavior in ferroelectric capacitors. *Appl. Phys. Lett.*, 82(18):3071, 2003.

[40] N. A. Basit and H. K. Kim. Growth of highly oriented Pb(Zr,Ti)O$_3$ films on MgO-buffered oxidized Si substrates and its application to ferroelectric nonvolatile memory field-effect transistors. *Appl. Phys.*, 73(26):3941, 1998.

[41] K. R. Bellur, H. N. Al-Shareef, S. H. Rou, K. D. Gifford, O. Auciello, and A. I. Kingon. Electrical Characterization of Sol Gel Derived PZT Thin Films. *Int. Symp. Appl. Ferroelectrics*, Proc. 8th IEEE:448, 1992.

[42] D. Akai, K. Hirabayashi, M. Yokawa, K. Sawada, and M. Ishida. Epitaxial growth of Pt(001) thin films on Si substrates using an epitaxial γ-Al$_2$O$_3$ (001) buffer layer. *J. Cryst. Growth*, 264:463, 2004.

[43] M. Alexe and J. F. Scott. Nanophase ferroelectric ceramic memories. *Key Eng. Mater.*, 206:1267, 2002.

[44] M. Alexe, C. Harnagaa, D. Hesse, and U. Gösele. Patterning and switching of nanosize ferroelectric memory cells. *Appl. Phys. Lett.*, 75(12):1793, 1999.

[45] M. Alexe, C. Harnagea, D. Hesse, and U. Gösele. Polarization imprint and size effect in mesoscopic ferroelectric structures. *Appl. Phys. Lett.*, 79(2):242, 2001.

[46] M. Alexe, C. Harnagea, W. Erfurth, D. Hesse, and U. Gösele. 100-nm lateral size ferroelectric memory cells fabricated by electron-beam direct writing. *Appl. Phys. A*, 70:247, 2000.

[47] H. G. Craighead and L. M. Schiavone. Metal deposition by electron beam exposure of an organometallic film. *Appl. Phys. Lett.*, 48(25):1748, 1986.

[48] J. Lohau, S. Friedrichowski, G. Dumpich, and E. F. Wassermann. Electron-beam lithography with metal colloids: Direct writing of metallic nanostructures. *J. Vac. Sci. Technol. B*, 16(1):77, 1998.

[49] S. Tiedke, T. Schmitz, K. Prume, A. Roelofs, T. Schneller, U. Kall, R. Waser, C. S. Ganpule, V. Nagarajan, A. Stanishevsky, and R. Ramesh. Direct hysteresis measurements of single nanosized ferroelectric capacitors contacted with an atomic force microscope. *Appl. Phys. Lett.*, 79(22):3678, 2001.

[50] A. Stanishevsky, S. Aggarwal, A. S. Prakash, J. Melnagilis, and R. Ramesh. Focused ion-beam patterning of nanoscale ferroelectric capacitors. *J. Vac. Sci. Technol. B*, 16(6):3899, 1998.

[51] A. L. Roytburd, S. P. Alpay, V. Nagarajan, C. S. Ganpule, and S. Aggarwal. Measurement of internal stresses via the polarization in epitaxial ferroelectric fiilms. *Phys. Rev. Lett.*, 85(1):190, 2000.

[52] C. Harnagea, M. Alexe, J. Schilling, J. Choi, R. Wehrspohn, D. Hesse, and U. Gösele. Mesoscopic ferroelectric cell arrays prepared by imprint lithography. *Appl. Phys. Lett.*, 83(9):1827, 2003.

[53] J.-S. Lee, B.-I. Lee, and S.-K. Joo. Single-grained PZT thin films for high level FRAM integration-fabrication and characterization. *Integrated Ferroelectrics*, 31:149, 2000.

[54] J.-S. Lee, E. Park, J. Ho, B. Lee, and S.-K. Loo. A study on the selective nucleation for formation of large single grains in PZT thin films. *MRS Symp. Proc.*, 596:217, 2000.

[55] J. Lee and S. Joo. Self-limiting behavior of the grain growth in lead zirconate titanate thin films. *J. Appl. Phys.*, 92(5):2658, 2002.

[56] H. Tang, J. A. Bardwell, J. B. Webb, S. Moisa, J. Fraser, and S. Rolfe. Selective growth of GaN on a SiC substrate patterned with an AlN seed layer by ammonia molecular-beam epitaxy. *Appl. Phys. Lett.*, 79(17):2764, 2001.

[57] P. Muralt, S. Hiboux, and M. Cantoni. Seed Layers of the titania - lead oxide system. *Mat. Res. Soc. Symp. Proc.*, 748:75, 2003.

[58] M. Shimizu, H. Fujisawa, H. Niu, and K. Honda. Growth of ferroelectric $PbZr_xTi_{1-x}O_3$ thin films by metalorganic chemical vapor deposition (MOCVD). *J. Cryst. Growth*, 237-239:448, 2002.

[59] A. Roelofs, T. Schneller, K. Szot, and R. Waser. Towards the limit of ferroelectric nanosized grains. *Nanotechnology*, 14:250, 2003.

[60] I. Szafraniak, C. Harnagea, R. Scholz, S. Bhattacharyya, D. Hesse, and M. Alexe. Ferroelectric epitaxial nanocrystals obtained by a self-patterning method. *Appl. Phys. Lett.*, 83(11):2211, 2003.

[61] M. Alexe, J. F. Scott, C. Curran, N. D. Zakharov, D. Hesse, and A. Pignolet. Self-patterning nano-electrodes on ferroelectric thin films for gigabit memory applications. *Appl. Phys. Lett.*, 73(11):1592, 1998.

[62] M. G. Alexe, A. Gruverman, C. Harnagea, N. D. Zakharov, A. Pignolet, D. Hesse, and J. F. Scott. Switching properties of self-assembled ferroelectric memory cells. *Appl. Phys. Lett.*, 75(8):1158, 1999.

[63] C. Harnagea, A. Pignolet, M. Alexe, D. Hesse, and U. Gösele. Quantitative ferroelectric characterization of single submicron grains in Bi-layered perovskite thin films. *Appl. Phys. A*, 70:261, 2000.

[64] Y. Luo, I. Szafraniak, N. Zakharov, V. Nagarajan, M Steinfart, R. Wehrspohn, J. Wendorff, R. Ramesh, and M. Alexe. Nanoshell tubes of ferroelectric lead zirconate titanate and barium titanate. *Appl. Phys. Lett.*, 83(3):440, 2003.

[65] B. M. Jin, J. Kim, and S. C. Kim. Effects of grain size on the electrical properties of $PbZr_{0.52}Ti_{0.48}O_3$ ceramics. *Appl. Phys A.*, 65:53, 1997.

[66] G. Arlt. Microstructure and domain effects in ferroelectric ceramics. *Ferroelectrics*, 91:3, 1989.

[67] M. H. Frey and D. A. Payne. Grain-size effect on structure and phase transformations for barium titanate. *Phys. Rev. B*, 54(5):3158, 1996.

[68] G. Arlt and H. Peusens. The dielectric constant of coarse grained $BaTiO_3$ ceramics. *Ferroelectrics*, 48:213, 1983.

[69] J. F. Meng, R. S. Katiyar, and G. T. Zou. Grain size effect on ferroelectric phase transition in $Pb_{1-x}Ba_xTiO_3$. *J. Phys. Chem. Solids*, 59(6-7):1161, 1998.

[70] Y. Park and K. Knowles. Particle-size effect on the ferroelectric phase transition in $PbSc_{0.5}Ta_{0.5}O_3$ ceramics. *J. Appl. Phys.*, 83(11):5702, 1998.

[71] P. Würfel and I. P. Batra. Depolarizaton effects in thin ferroelectric films. *Ferroelectrics*, 12:55, 1976.

[72] S. Li, J. A. Eastman, Z. Li, C. M. Foster, R. E. Newnham, and L. E. Cross. Size effects in nanostructured ferroelectrics. *Phys. Lett. A*, 212:341, 1996.

[73] C. Jaccard, W. Känzig, and M. Peter. Das Verhalten von kolloidalen Seignetteelektrika I, Kaliumphosphat KH_2PO_4. *Helv. Phys. Acta*, 26:521, 1953.

[74] O. G. Vendik, S. P. Zubko, and L. T. Ter-Martirosayn. Experimental evidence of the size effect in thin ferroelectric films. *Appl. Phys. Lett.*, 73(1):37, 1998.

[75] A. M. Glass and J. W. Shiever. Evolution of ferroelectricity in ultrafine-grained $Pb_5Ge_3O_{11}$ crystallized from the glass. *J. Appl. Phys.*, 48(12):5213, 1977.

[76] S. H. Wemple. Polarization Fluctuations and the Optical-Absorption Edge in $BaTiO_3$. *Phys. Rev. B*, 2(7):2679, 1970.

[77] W. L. Warren, L. Robertson, D. Dimos, B. A. Tuttle, and D. A. Pyne. Pb displacements in $Pb(Ti, Zr)O_3$ perovskites. *Phys. Rev. B*, 53(6):3080, 1996.

[78] U. V. Waghmare and K. M. Rabe. Ab initio statistical mechanics of the ferroelectric phase transition in $PbTiO_3$. *Phys. Rev. B*, 55(10):6161, 1997.

[79] P. Ghosez and K. M. Rabe. Microscopic model of ferroelectricity in stress-free $PbTiO_3$ ultrathin films. *Appl. Phys. Lett.*, 76(19):2767, 2000.

[80] K. M. Rabe and P. Ghosez. Ferroelectricity in $PbTiO_3$ thin films: A first principles approach. *J. Elecroceram.*, 4(2/3):379, 2000.

[81] T. Tybell, C. H. Ahn, and J.-M. Triscone. Ferroelectricity in thin perovskite films. *Appl. Phys. Lett.*, 75(6):856, 1999.

[82] D. R. Tilley. *Ferroelectric Ceramics*. Phase Trans. Birkhäuser, Basel, 1993.

[83] L. Landau and E. Lifchitz. *Electrodynamique des Milieux Continus*, volume 8 of *Physique Théorique*. Moscou, mir edition, 1969.

[84] P. W. Forsbergh. Effect of two-dimensional pressure on the curie point of barium titanate. *Phys. Rev.*, 93(4):686, 1954.

[85] N. A. Pertsev, V. G. Kukhar, H. Kohlstedt, and R. Waser. Phase diagrams and physical properties of single-domain epitaxial $Pb(Zr_{1-x}Ti_x)O_3$ thin films. *Phys. Rev. B*, 67(054107), 2003.

[86] C. Wang and S. Smith. The size effect on the switching properties of ferroelectric films: A one-dimensional model. *J. Phys. Condens. Mat.*, 8(26):4813, 1996.

[87] L. Chen, V. Nagarajan, R. Ramesh, and A. L. Roytburd. Nonlinear electric field dependence of piezoresponse in epitaxial ferroelectric lead zirconate titanate thin films. *J. Appl. Phys.*, 94(8):5147, 2003.

[88] T. Mitsui, I. Tatsuzaki, and E. Nakamira. *An introduction to the physics of ferroelectrics*. Gordon and Breach Science Publishers, New York, 1974.

[89] W. J. Merz. Double hysteresis loop of $BaTiO_3$ at the Curie point. *Phys. Rev.*, 91(3):513, 1953.

[90] V. A. Zhirnov. A contribution to the theory of domain walls in ferroelectrics. *Sov. Phys. JETP-USSR*, 8(5):822, 1959.

[91] J. P. Remeika and A. M. Glass. Growth and ferroelectric properties of high resistivity single crystals of lead titanate. *MRS Bulletin*, 5(1):37, 1970.

[92] M. J. Haun, Z. Q. Zhuang, E. Furman, S. J. Jang, and L. E. Cross. Thermodynamic theory of the lead zirconate-titanate solid-solution system, 3. Curie constant and 6th-order polarization interaction dielectric stiffness coefficients. *Ferroelectrics*, 99:45, 1989.

[93] I. P. Batra, P. Wurfel, and B. D. Silverman. Phase Transition, Stability, and Depolarization Field in Ferroelectric Thin Films. *Phys. Rev. B*, 8(7):3257, 1973.

[94] K. Binder. Surface effects on phase transitions in ferroelectrics and antiferroelectrics. *Ferroelectrics*, 35:99, 1981.

[95] D. T. Tilley. Phase transformation in ferroelectric films. *Ferroelectrics*, 134:313, 1992.

[96] J. F. Scott, H. M. Duiker, D. Beale, P. Poulighy, K. Dimmler, M. Parris, D. Butler, and S. Eaton. Properties of ceramic KNO_3 thin-film memories. *Physica B*, 150:160, 1988.

[97] M. Anliker, H. R. Brugger, and W. Känzig. Das Verhalten von kolloidalen Seignetteelektrika III, Bariumtitanat $BaTiO_3$. *Helv. Phys. Acta*, 27:99, 1954.

[98] I. P. Batra and B. D. Silverman. Thermodynamic stability of thin ferroelectric films. *Sol. St. Comm.*, 11:291, 1972.

[99] I. P. Batra, P. Wurfel, and B. D. Silverman. New Type of Phase Transition in Ferroelectric Thin Films. *Phys. Rev. Lett.*, 30(9):384, 1973.

[100] P. Wurfel and I. P. Batra. Depolarization-field-induced instability in thin ferroelectric films-experiment and theory. *Phys. Rev. B*, 8(11):5126, 1973.

[101] W. Y. Shih, W.-H. Shih, and I. A. Aksay. Size dependence of ferroelectric transition of small $BaTiO_3$ particles: Effect of depolarization. *Phys. Rev. B*, 50(21):15575, 1994.

[102] S. Triebwasser. Space Charge Fields in $BaTiO_3$. *Phys. Rev.*, 118(1):100, 1960.

[103] W. Känzig. Space charge layer near the surface of a ferroelectric. *Phys. Rev.*, 98:549, 1955.

[104] Y. G. Wang, W. L. Zhong, and P. L. Zhang. Surface and size effects on ferroelectric films with domain structures. *Phys. Rev. B*, 51(8):5311, 1995.

[105] D. R. Tilley and B. Zeks. Landau theory of phase transitions in thick films. *Sol. St. Comm.*, 49(8):823, 1984.

[106] W. L. Zhong, Y. G. Wang, P. L. Zhang, and B. D. Qu. Phenomenological study of the size effect on phase transitions in ferroelectric particles. *Phys. Rev. B*, 50(2):698, 1994.

[107] R. Kretschmer and K. Binder. Surface effects on phase transitions in ferroelectrics and dipolar magnets. *Phys. Rev. B*, 20(3):1065, 1979.

[108] M. Abplanalp. *Piezoresponse scanning force microscopy of ferroelectric domains*. PhD thesis, Swiss Federal Institute of Technology, 1970.

[109] S. Li, J. A. Eastman, J. M. Vetrone, C. M. Foster, R. E. Newham, and L. E. Cross. Dimension and size effects in ferroelectrics. *Jpn. J. Appl. Phys.*, 36:5169, 1997.

[110] H. Huang, C. Sun, Z. Tianshu, and P. Hing. Grain-size effect on ferroelectric $Pb(Zr_{1-x}Ti_x)O_3$ solid solutions induced by surface bond contraction. *Phys. Rev. B*, 63:184112, 2001.

[111] H. Huang, C. Q. Sun, and P. Hing. Surface bond contraction and its effect on the nanometric sized lead zirconate titanate. *J. Phys. Condens. Mat.*, 12:L127, 2000.

[112] M. de Keijser, G. J. M. Dormans, P. J. van Veldhoven, and D. M. Leeuw. Effects of crystallite size in PbTiO$_3$ thin films. *Appl. Phys. Lett.*, 59(27):3556, 1991.

[113] K. Ishikawa, K. Yoshikawa, and N. Okada. Size effect on the ferroelectric phase transition in PbTiO$_3$ ultrafine particles. *Phys. Rev. B*, 37(10):5852, 1988.

[114] S. Chattopadhyay, P. Ayyub, V. R. Palkar, and M. Multani. Size-induced phase transition in the nanocrystalline ferroelectric PbTiO$_3$. *Phys. Rev. B*, 52(18):13177, 1995.

[115] I. Rychetsky and O. Hudak. The ferroelectric phase transition in small spherical particles. *J. Phys. Condens. Mat.*, 9:4955, 1997.

[116] K.-H. Chew, J. Osman, D. R. Tilley, F. G. Shin, and H. L. W. Chan. Surface aided polarization reversal in small ferroelectric particles. *J. Appl. Phys.*, 93(7):4215, 2003.

[117] K. Ishikawa, T. Nomura, N. Okada, and K. Takada. Size effect on the phase transition in PbTiO$_3$ fine particles. *Jpn. J. Appl. Phys.*, 35(Part 1, 9B):5196, 1996.

[118] J. F. Scott. Phase transitions in ferroelectric thin films. *Phase Trans.*, 30:107, 1991.

[119] F. Yan, P. Bao, H. Chan, C. Choy, and Y. Wang. The grain size effect of Pb(Zr$_{0.3}$Ti$_{0.7}$O$_3$) thin films. *Thin Solid Films*, 406:282, 2002.

[120] P. D. Beale and H. M. Duiker. Grain-size effects in ferroelectric switching. *Ferroelectrics*, 117:165, 1991.

[121] C. J. Lu, H. M. Shen, S. B. Ren, and Y. N. Wang. X-ray evidence for the absence of 90°-domains in polycrystalline PbTiO$_3$ thin films with grains below a critical size. *Appl. Phys. A*, 65:395, 1997.

[122] L. Zhang, W. L. Zhong, C. L. Wang, Y. P. Peng, and Y. G. Wang. Size dependence of dielectric properties and structural metastability in ferroelectrics. *Eur. Phys. J. B*, 11:565, 1999.

[123] S. B. Desu. Suppression of Size Effects in Ferroelectric Films. *MRS Symp. Proc.*, 541:457, 1999.

[124] Y. Ishibashi and H. Orihara. Size effect in ferroelectric switching. *J. Phys. Soc. Jpn.*, 61(12):4650, 1992.

[125] W. J. Merz. Domain Formation and Domain Wall Motions in Ferroelectric BaTiO$_3$ Single Crystals. *Phys. Rev.*, 95(3):690, 1954.

[126] R. S. Katiyar and J. F. Meng. Grain size effect on Ferroelectric phase transitions in nano-crystalline perovskite materials. *Ferroelectrics*, 229:201, 1999.

[127] W. L. Zhong, B. Jiang, P. L. Zhang, J. M. Ma, H. M. Cheng, Z. H. Yang, and L. X. Li. Phase transition in PbTiO$_3$ ultrafine particles of different sizes. *J. Phys. Condens. Mat.*, 5:2619, 1993.

[128] I. Yamashita, H. Kawaji, and T. Atake. Particle-size effects on the III-IV phase transition in CsZnPO$_4$. *Phys. Rev. B*, 68(9):092104, 2003.

[129] M. Hamada, H. Tabata, and T. Kawai. Size effect of the dielectric properties in bismuth-based layer-structured ferroelectric films. *Jpn. J. Appl. Phys.*, 37(Part 1, 9B):5174, 1998.

BIBLIOGRAPHY 159

[130] V. P. Dudkevich, V. A. Bukreev, V. M. Mukhortov, Y. I. Golovko, and Y. G. Sindeev. Internal size effect in concensed $BaTiO_3$ ferroelectric films. *Phys. Stat. Sol. (A)*, 65:463, 1981.

[131] T. Yu, Z. Shen, W. Toh, J. Xue, and J. Wang. Size effect on the ferroelectric phase transition in $SrBi_2Ta_2O_9$ nanoparticles. *J. Appl. Phys.*, 94(1):618, 2003.

[132] B. Jiang, J. L. Peng, L. A. Bursill, and W. L. Zhong. Size effects on ferroelectricity of ultrafine particles of $PbTiO_3$. *J. Appl. Phys.*, 87(7):3462, 2000.

[133] W. R. Bussem, L. E. Cross, and A. K. Goswami. Phenomenolocigal theory of high permittivity in fine-grained barium titanate. *J. Am. Ceram. Soc.*, 49(1):33, 1966.

[134] W. R. Bussem, L. E. Cross, and A. K. Goswami. Effect of two-dimensional pressure on the permittivity of fine- and coarse-grained barium titanate. *J. Am. Ceram. Soc.*, 49(1):36, 1966.

[135] T. Mitsui and J. Furuichi. Domain structure of rochelle salt and KH_2PO_4. *Phys. Rev.*, 90(2):193, 1953.

[136] Z. Zhang and M. G. Legally. Atomistic processes in the early stages of thin-film growth. *Science*, 276:377, 1997.

[137] B. Lewis and D. S. Campbell. Nucleation and initial growth-behavior of thin-film deposits. *J. Vac. Sci. Technol.*, 4(5):209, 1967.

[138] A. Hadni and R. Thomas. The use of regular distribution of minute pinholes for the epitaxial growth of an oriented thin film. *Thin Solid Films*, 81:247, 1981.

[139] A. Hadni, T. Thomas, and C. Erhard. An unusual type of epitaxial growth. *Phys. Stat. Sol. (A)*, 39:419, 1977.

[140] N. Motta, F. Rosei, A. Sgarlata, G. Capellini, S. Mobilio, and F. Boscherini. Evolution of the intermixing process in Ge/Si(111) self-assembled islands. *Mater. Sci. Eng. B*, 88:264, 2002.

[141] B. Lewis and J. C. Anderson. *Nucleation and growth of thin films*. Academic Press Inc., London, 1978.

[142] J. Ziolkowski. New method of calculation of the surface enthaply of solids. *Surf. Sci.*, 209:536, 1989.

[143] T. Ning, Q. Yu, and Y. Ye. Multilayer relaxaition at the surface of fcc metals: Cu, Ag, Au, Ni, Pd, Pt, Al. *Surf. Sci.*, 206:L857, 1988.

[144] W. R. Tyson and W. A. Miller. Surface free energies of solid metals. Estimation from liquid surface tension measurements. *Surf. Sci.*, 62:267, 1977.

[145] J. W. Matthews. Epitaxial growth. *Materials Science and Technology Series*, 1(Part B):429, 1975.

[146] J. H. van der Merwe. Crystal Interfaces. Part I. Semi-infinite Crystals. *J. Appl. Phys.*, 34:117, 1963.

[147] J. W. Matthews. *Dislocations in Solids*, volume 2. North-Holland, 1979.

[148] G. Gutekunst, J. Mayer, and M. Rühle. Atomic structure of epitaxial Nb-Al$_2$O$_3$ interfaces I. Coherent regions. *Philosoph. Mag. A*, 75(5):1329, 1997.

[149] M. W. Finnis. The theory of metal-ceramic interfaces. *J. Phys. Condens. Mat.*, 8:5811, 1996.

[150] J. W. Matthews and A. E. Blakeslee. Defects in epitaxial multilayers I. Misfit dislocations. *J. Cryst. Growth*, 27:118, 1974.

[151] G. Gutekunst, J. Mayer, and M. Rühle. Atomic structure of epitaxial Nb-Al$_2$O$_3$ interfaces II. Misfit dislocations. *Philosoph. Mag. A*, 75(5):1329, 1997.

[152] A. E. Romanov, T. Wagner, and M. Rühle. Coherent to incoherent transition in mismatched interfaces. *Scripta Mater.*, 38(6):869, 1997.

[153] L. B. Freund. Dislocation mechanisms of relaxation in strained epitaxial films. *MRS Bulletin*, 17(7):52, 1992.

[154] J. W. Matthews and A. E. Blakeslee. Defects in epitaxial multilayers II. Dislocation pile-ups, threading dislocations, slit lines and cracks. *J. Cryst. Growth*, 29:273, 1975.

[155] J. W. Matthews and A. E. Blakeslee. Defects in epitaxial multilayers III. Preparation of almost perfect multilayers. *J. Cryst. Growth*, 32:265, 1976.

[156] U. Schönberger, O. K. Andersen, and M. Methfessel. Bonding at metal-ceramic interfaces: Ab initio density-functional calculations for Ti and Ag on MgO. *Acta Met. Mat.*, 40:S1, 1992.

[157] C. Kruse, M. W. Finnis, J. S. Lin, M. C. Payne, V. Milman, A. de Vita, and M. J. Gillan. First-principle study of the atomistic and electronic structure of the niobium-α-alumina (0001) interface. *Philosoph. Mag. Lett.*, 73(6):377, 1996.

[158] T. Wagner, A. D. Polli, G. Richter, and H. Stanzick. Epitaxial growth of metals on (100) SrTiO$_3$: The influence of lattice mismatch and reactivity. *Z. Metallkd.*, 92:701, 2001.

[159] S. Köstlmeier and C. Elsässer. Ab-Initio inverstigation of metal-ceramic bonding: M(001)/MgAl$_2$O$_4$ (001), M=Al, Ag. *Interf. Sci.*, 8:41, 2000.

[160] R. Schweinfest, S. Köstlmeier, F. Ernst, C. Elsässer, and T. Wagner. Atomistic and electronic structure of Al/MgAl$_2$O$_4$ and Ag/MgAl$_2$O$_4$ interfaces. *Philosoph. Mag. A*, 81(4):927, 2001.

[161] R. Benedek, A. Alavi, D. N. Seidman, D. A. Muller, and C. Woodward. First principle simulation of a ceramic/metal interface with misfit. *Phys. Rev. Lett.*, 84(15):3362, 2000.

[162] R. Benedek and D. N. Seidman. Atomic and elecronic structure and interatomic potentials at a polar ceramic/metal interface: {222}MgO/Cu. *Phys. Rev. B*, 60(23):16094, 1999.

[163] P. Alemany. Metal-ceramic adhesion: Band structure calculations on transition-metal-AlN interfaces. *Surf. Sci.*, 314:114, 1994.

[164] A. Seifert, F. F. Lange, and J. S. Speck. Epitaxial growth of PbTiO$_3$ thin films on (001) SrTiO$_3$ from solution precursors. *J. Mater. Res.*, 10(3):680, 1995.

[165] M. Wagner, T. Wagner, D. L. Carroll, J. Marien, D. A. Bonnell, and M. Rühle. Model systems for metal-ceramic interface studies. *MRS Bulletin*, 22(8):42, 1997.

[166] R. J. Needs and M. Mansfield. Calculatinos of the surface stress tensor and surface energy of the (111) surfaces of iridium, platinum and gold. *J. Phys. Condens. Mat.*, 1:7555, 1989.

[167] A. Wachter, K. P. Bohnen, and K. M. Ho. Structure and dynamics at the Pt (100)-surface. *Ann. Physik*, 5:215, 1996.

[168] M. I. Haftel. Surafce reconstruction of platinum and gold and the embedded-atom model. *Phys. Rev. B*, 48(4):2611, 1993.

[169] E. Gehrer and K. Hayek. Ultrathin epitaxial deposits of platinum on NaCl to be used as model catalysts. *Thin Solid Films*, 115:283, 1984.

[170] K. Zhao and H. K. Wong. Epitaxial growth of platinum thin films on various substrates by facing-target sputtering technique. *J. Cryst. Growth*, 256:283, 2003.

[171] A. D. Polli, T. Wagner, T. Gemming, and M. Rühle. Growth of platinum on TiO_2- and SrO-terminated $SrTiO_3$ (100). *Surf. Sci.*, 448:279, 2001.

[172] K. Reichelt, T. Schober, and J. Viehweg. On the growth of thick eptiaxial platinum films with various orientations. *J. Cryst. Growth*, 18:312, 1973.

[173] J. Koshy. Epitaxial growth and structure of evaporated platinum films on rock salt. *J. Cryst. Growth*, 34:345, 1976.

[174] J. W. Matthews. Growth of face-centered-cubic metals on sodium chloride substrates. *J. Vac. Sci. Technol.*, 3:133, 1965.

[175] J. Dath, P. Deschamps, J. Leleux, and J. Cauchot. Growth of monocrystalline platinum thin films by vacuum deposition on LiF (100) under controlled electron bombardement. *Thin Solid Films*, 131:31, 1985.

[176] R. E. Leuchtner, D. Chrisey, J. S. Horwitz, and K. S. Grabowski. The preparation of epitaxial platinum films by pulsed laser deposition. *Surf. Coatings Technol.*, 51:476, 1992.

[177] P. C. McIntyre, C. J. Maggiore, and M. Nastasi. Epitaxy of Pt thin films on (001) MgO-II: Orientation evolution from nucleation through coalescence. *Acta Mater.*, 45(2):879, 1997.

[178] K. H. Ahn and S. Baik. Change of growth orientation in Pt films epitaxially grown on MgO (001) substrates by sputtering. *J. Mater. Res.*, 17(9):2334, 2002.

[179] S. Horita, S. Sasaki, O. Kitagawa, and S. Horii. Ferroelectric properties of epitaxial $Ba_4Ti_3O_{12}$ films deposited on epitaxial (001) Ir and (100) Pt films on Si by sputtering. *Vacuum*, 66:427, 2002.

[180] T. Ishikawa, Y. Abe, S. Shinaki, and K. Sasaki. Epitaxial Ir thin film on (001) MgO single crystal prepared by sputtering. *Jpn. J. Appl. Phys.*, 42(Part 1, 9A):5747, 2003.

[181] K. Iijima, Y. Tomita, R. Takayama, and I. Ueda. Preparation of c-axis oriented $PbTiO_3$ thin films and their crystallographic, dielectric, and pyroelectric properties. *J. Appl. Phys.*, 60(1):361, 1986.

[182] K. D. Budd, S. K. Dey, and D. A. Payne. Sol-gel processing of PbTiO$_3$, PbZrO$_3$, PZT, and PLZT thin films. *Br. Ceram. Proc.*, 36:107, 1985.

[183] C. M. Foster, G.-R. Bai, R. Csencsits, J. Vetrone, R. Jammy, L. A. Wills, E. Carr, and J. Amano. Single-crystal Pb(Zr$_x$Ti$_{(1-x)}$)O$_3$ thin films prepared by metal-organic chemical vapor deposition: Systematic compositional variation of electronic and optical properties. *J. Appl. Phys.*, 81(5):2349, 1997.

[184] K. Nashimoto and S. Nakamura. Preparation and Characterization of Sol-Gel Derived Epitaxial and Oriented Pb(Zr$_{0.52}$Ti$_{0.48}$)O$_3$ Thin Films. *Jpn. J. Appl. Phys.*, 33(Part 1, 9B):5147, 1994.

[185] E. Tokumitsu, K. Itani, B.-K. Moon, and H. Ishiwara. Crystalline Quality and Electrical Properties of PbZr$_x$Ti$_{(1-x)}$O$_3$ Thin Films Prepared on SrTiO$_3$-covered Si Substrates. *Jpn. J. Appl. Phys.*, 34:5202, 1995.

[186] C. R. Cho, L. F. Francis, and D. L. Polla. Ferroelectric properties of sol-gel deposited Pb(Zr,Ti)O$_3$/LANiO$_3$ thin films on single crystal and platinized-Si substrates. *Mater. Lett.*, 38:125, 1999.

[187] J. S. Horwitz, K. S. Grabowski, D. B. Chrisey, and R. E. Leuchtner. In situ deposition of epitaxial PbZr$_x$Ti$_{(1-x)}$O$_3$ thin films by pulsed laser deposition. *Appl. Phys. Lett.*, 59(13):1565, 1991.

[188] K. Nashimoto, D. K. Fork, and G. B. Anderson. Solid phase epitaxial growth of sol-gel derived Pb(Zr,Ti)O$_3$ thin films on SrTiO$_3$ and MgO. *Appl. Phys. Lett.*, 66(7):822, 1995.

[189] M. de Keijser, G. J. M. Dormans, J. F. M. Cillessen, and D. M. de Leeuw. Epitaxial PbTiO$_3$ thin films grown by organometallic chemical vapor deposition. *Appl. Phys. Lett.*, 58(23):2636, 1991.

[190] B. S. Kwak, E. P. Boyd, and A. Erbil. Metalorganic chemical vapor deposition of PbTiO$_3$ thin films. *Appl. Phys. Lett.*, 53(18):1702, 1988.

[191] G. R. Bai, H. L. M. Chang, C. M. Foster, Z. Shen, and D. J. Lam. The relationship between the MOCVD parameters and the crystallinity, epitaxy, and domain structure of PbTiO$_3$ films. *J. Mater. Res.*, 9(1):156, 1994.

[192] K. S. Lee and S. Baik. Reciprocal space mapping of phase transformation in epitaxial PbTiO$_3$ thin films using synchrotron X-ray diffraction. *J. Appl. Phys.*, 85(3):1995, 1999.

[193] W. Braun, B. S. Kwak, A. Erbil, J. D. Budai, and B. J. Wilkens. Epitaxial lead zirconate-titanate thin films on sapphire. *Appl. Phys. Lett.*, 63(4):467, 1993.

[194] Y. K. Kim, K. Lee, and S. Baik. Domain structure of epitaxial PbTiO$_3$ thin films on Pt(001)/MgO(001) substrates. *J. Appl. Phys.*, 95(1):236, 2004.

[195] R. A. McKee, F. J. Walker, and M. F. Chisholm. Crystalline Oxides on Silicon: The First Five Monolayers. *Phys. Rev. Lett.*, 81(14):3014, 1998.

[196] R. S. Batzer, B. M Yen, D. Liu, and H. chen. A high-temperature X-ray-diffraction study of epitaxial PbTiO$_3$ thin films on MgO(100) grown by metal-organic chemical-vapor deposition. *J. Appl. Phys.*, 80(11):6235, 1996.

BIBLIOGRAPHY 163

[197] Y. Gao, G. Bai, K. L. Merkle, Y. Shi, H. L. M. Chang, Y. Shen, and D. J. Lam. Microstructure of PbTiO$_3$ thin films deposited on (001) MgO by MOCVD. *J. Mater. Res.*, 8(1):145, 1993.

[198] S. P. Alpay, V. Nagarajan, L. A. Bendersky, M. D. Vaudin, S. Aggarwal, R. Ramesh, and A. L. Roytburd. Effect of the electrode layer on the polydomain structure of epitaxial PbZr$_{0.2}$Ti$_{0.8}$O$_3$ thin films. *Appl. Phys.*, 85(6):3271, 1999.

[199] R. E. Leuchtner, K. S. Grabowski, D. B. Chrisey, and J. S. Horwitz. Anion-assisted pulsed laser deposition of lead zirconate titanate films. *Appl. Phys. Lett.*, 60(10):1193, 1992.

[200] J. Lee, A. Safari, and R. L. Pfeffer. Growth of epitaxial Pb(Zr,Ti)O$_3$ films by pulsed laser deposition. *Appl. Phys. Lett.*, 61(14):1643, 1992.

[201] K. S. Lee and S. Baik. Thickness dependence of domain formation in epitaxial PbTiO$_3$ thin films grown on MgO (001) substrates. *J. Appl. Phys.*, 87(11):8035, 2000.

[202] R. Takayama and Y. Tomita. Preparation of epitaxial Pb(Zr$_x$Ti$_{1-x}$)O$_3$ thin films and their crystallographic, pyroelectric, and ferroelectric properties. *J. Appl. Phys.*, 65(4):1666, 1989.

[203] K. Iijima, I. Ueda, and K. Kugimiya. Preparation and Properties of Lead Zirconate-Titanate Thin Films. *Jpn. J. Appl. Phys.*, 30(9B):2149, 1991.

[204] C. J. Lu, H. M. Shen, and Y. N. Wang. Preparation and crystallization of Pb(Zr$_{0.95}$Ti$_{0.05}$)O$_3$ thin films deposited by radio-frequency magnetron sputtering with a stoichiometric ceramic target. *Appl. Phys. A*, 67:253, 1998.

[205] T. Maeder, P. Muralt, and L. Sagalowicz. Growth of (111)-oriented PZT on RuO$_2$ (100)/Pt (111) electrodes by in-situ sputtering. *Thin Solid Films*, 345:300, 1999.

[206] C. M. Foster, Z. Li, M. Buckett, D. Miller, P. M. Baldo, L. E. Rehn, G. R. Bai, D. Guo, H. You, and K. L. Merkle. Substrate effects on the structure of epitaxial PbTiO$_3$ thin films prepared on MgO, LaAlO$_3$, and SrTiO$_3$ by metalorganic chemical-vapor deposition. *J. Appl. Phys.*, 78(4):2607, 1995.

[207] G. J. M. Dormans, P. J. van Veldhoven, and M. de Keijser. Composition-controlled growth of PbTiO$_3$ on SrTiO$_3$ by organometallic chemical vapour deposition. *J. Cryst. Growth*, 123:537, 1992.

[208] A. K. Sharma, J. Narayan, C. Jin, A. Kvit, S. Chattopadhyay, and C. Lee. Integration of Pb(Zr$_{0.52}$Ti$_{0.48}$)O$_3$ epilayers with Si by domain epitaxy. *Appl. Phys. Lett.*, 76(11):1458, 2000.

[209] C. M. Foster, G.-R. Bai, Z. Li., and R. Jammy. Properties Variation with Composition of Single-Crystal Pb(Zr$_x$Ti$_{1-x}$)O$_3$ Thin Films Prepared by MOCVD. *MRS Symp. Proc.*, 401:139, 1996.

[210] S. Hiboux, P. Muralt, and T. Maeder. Domain and lattice contributions to dielectric and piezoelectric properties of Pb(Zr$_x$,Ti$_{1-x}$)O$_3$ thin films as a function of composition. *J. Mater. Res.*, 14(11):4307, 1999.

[211] G. Arlt and N. A. Pertsev. Force constant and effective mass of 90° domain walls in ferroelectric ceramics. *J. Appl. Phys.*, 70(4):2283, 1991.

[212] W. Pompe, X. Gong, Z. Suo, and J. S. Speck. Elastic energy release due to domain formation in the strained epitaxy of ferroelectric and ferroelastic films. *J. Appl. Phys.*, 74(10):6012, 1993.

[213] A. L. Roitburd. Equilibrium Structure of Epitaxial Layers. *Phys. Stat. Sol. (a)*, 37:329, 1976.

[214] A. L. Roytburd. Thermodynamics of polydomain heterostructures. I. Effect of macrostresses. *J. Appl. Phys.*, 83(1):228, 1998.

[215] A. L. Roytburd. Thermodynamics of polydomain heterostructures. II. Effect of microstresses. *J. Appl. Phys.*, 83(1):239, 1998.

[216] S. P. Alpay and A. L. Roytburd. Thermodynamics of polydomain heterostructures. III. Domain stability map. *Appl. Phys.*, 83(9):4714, 1998.

[217] J. Lappalainen, J. Frantti, and V. Lantto. Electrical and mechanical properties of ferroelectric thin films laser ablated from a $Pb_{0.97}Nd_{0.02}(Zr_{0.55}Ti_{0.45})O_3$. *J. Appl. Phys.*, 82(7):3469, 1997.

[218] B. S. Kwak, A. Erbil, J. D. Budai, M. F. Chisholm, L. A. Boatner, and B. J. Wilkens. Domain formation and strain relaxation in epitaxial ferroelectric heterostructrures. *Phys. Rev. B*, 49(21):14865, 1994.

[219] S. H. Ling, Y. S. Tang, W. S. Au, and H. K. Wong. Epitaxial growth of $Pb(Zr,Ti)O_3$ films on $MgAl_2O_4$ by pulsed laser deposition. *Appl. Phys. Lett.*, 62(15):1757, 1993.

[220] V. Nagarajan, I. G. Jenkins, S. P. Alpay, H. Li, S. Aggarwal, L. Salamanca-Riba, A. L. Roytburd, and R. Ramesh. Thickness dependence of structural and electrical properties in epitaxial lead zirconate titanate films. *Appl. Phys.*, 86(1):595, 1999.

[221] S. P. Alpay, A. S. Prakash, S. Aggarwal, P. Shuk, M. Greenblatt, R. Ramesh, and A. L. Roytburd. Polydomain formation in epitaxial $PbTiO_3$ films. *Scripta Mater.*, 39(10):1435, 1998.

[222] W. J. Merz. Domain properties in $BaTiO_3$. *Phys. Rev.*, 88:421, 1952.

[223] C. S. Ganpule, V. Nagarajan, B. K. Hill, A. L. Roytburd, E. D. Williams, R. Ramesh, S. P. Alpay, and L. M. Eng. Imaging three-dimensional polarization in epitaxial polydomain ferroelectric thin films. *J. Appl. Phys.*, 91(3):1477, 2002.

[224] Z. Li, C. M. Foster, D. Guo, H. Zhang, G. R. Bai, P. M. Baldo, and L. E. Rehn. Growth of high quality single-domain single-crystal films of $PbTiO_3$. *Appl. Phys. Lett.*, 65(9):1106, 1994.

[225] W.-Y. Hsu and R. Raj. X-ray characterization of the domain structure of epitaxial lead titanate thin films on (001) strontium titanate. *Appl. Phys. Lett.*, 67(6):792, 1995.

[226] C. M. Foster, W. Pompe, A. C. Daykin, and J. S. Speck. Relative coherency strain and phase transformaiton history in epitaxial ferroelectric thin films. *J. Appl. Phys.*, 79(3):1405, 1996.

[227] C. Kittel. Theory of the structure of ferromagnetic domains in films and small particles. *Phys. Rev.*, 70(11/12):965, 1946.

BIBLIOGRAPHY

[228] S. B. Ren, C. J. Lu, H. M. Shen, and Y. N. Wang. In situ study of the evolution of domain structure in free-standing polycrystalline PbTiO$_3$ thin films under external stress. *Phys. Rev. B*, 55(6):3485, 1997.

[229] M. Copel, M. C. Reuter, E. Kaxiras, and R. M. Tromp. Surfactants in epitaxial growth. *Phys. Rev. Lett.*, 63(6):632, 1989.

[230] J. Camarero, J. Ferron, V. Cros, and L. Gomez. Atomistic mechanism of surfactant-assisted epitaxial growth. *Phys. Rev. Lett.*, 81(4):850, 1998.

[231] R. Tromp. Atoms go underground. *Nature Materials*, 2(4):212, 2003.

[232] P. Jang, S. Hong, and J. Kim. Role of Ag seed layer for CoCrPt/Ti perpendicular recording media. *J. Appl. Phys.*, 93(10):7741, 2003.

[233] C.-C. Leu, C.-H. Chien, M.-J. Yang, M.-C. Yang, T.-Y. Huang, H.-T. Lin, and C.-T. Hu. Effects of ultrathin tantalum seeding layers on sol-gel-derived SrBi$_2$Ta$_2$O$_9$ thin films. *Appl. Phys. Lett.*, 80(24):4600, 2002.

[234] Y. Gao, Y. Masuda, T. Yonezawa, and K. Koumoto. Site-selective deposition and micropatterning of SrTiO$_3$ thin film on self-assembled monolayers by the liquid phase deposition method. *Chem. Mater.*, 14:5006, 2002.

[235] N. Shirahata, Y. Masuda, T. Yonezawa, and K. Koumoto. Control over film thickness of SnO$_2$ ultrathin film selectively deposited on a patterned self-assembled monolayer. *Langmuir*, 18:10379, 2002.

[236] Y. Masuda, S. Ieda, and K. Koumoto. Site-selective depositon of Anatase TiO$_2$ in an Aqueous solution using a seed layer. *Langmuir*, 19:4415, 2003.

[237] M. Sayer and K. Sreenivas. Ceramic thin films: fabrication and applications. *Science*, 247(4946):1056, 1990.

[238] S. Hong, Y. Lee, J. Lee, and J. Kim. Growth behavior and ferroelectric properties of Zr-rich PZT thin films deposited on various Pt electrodes. *Integrated Ferroelectrics*, 23:65, 1999.

[239] K. Kushida-Abdelghafar, K. Torii, T. Mine, T. Kachi, and Y. Fujisaki. Orientation control in PZT/Pt/TiN multilayers with various Si and SiO$_2$ underlayers for high performance ferroelectric memories. *J. Vac. Sci. Technol.*, 18(1):231, 2000.

[240] G. Han, C. Yang, and J. Lee. Effect of (La,Sr)CoO$_3$ seed layer on the reliability of Pb(Zr,Ti)O$_3$ capacitors. *Integrated Ferroelectrics*, 25:341, 1999.

[241] U. Diebold, J.-M. Pan, and T. E. Madey. Ultrathin metal films on TiO$_2$ (110): Metal overlayer spreading and surface reactivity. *Surf. Sci.*, 287/288:896, 1993.

[242] R. D. Klissurska, T. Maeder, K. G. Brooks, and N. Setter. Microstructure of PZT sol-gel films on Pt substrates with different adhesion layers. *Microel. Eng.*, 29:297, 1995.

[243] K. Aoki, Y. Fukuda, K. Numata, and A. Nishimura. Effects of titanium layer on lead-zirconate-titanate crystallization processes in sol-gel deposition technique. *Jpn. J. Appl. Phys.*, 34:192, 1995.

[244] M. Adachi, T. Matsuzaki, T. Yamada, T. Shiosaki, and A. Kawabta. Sputter-Deposition of [111]-axis oriented rhombohedral PZT films and their dielectric, ferroelectric and pyro-electric properties. *Jap. J. Appl. Phys.*, 26(4):550, 1987.

[245] R. Thomas, S. Mochizuki, T. Mihara, and T. Ishida. Effects of $PbTiO_3$ seed layer on the characteristics of RF-sputtered $Pb(Zr_{0.5},Ti_{0.5}O_3)$ thin films. *Integrated Ferroelectrics*, 46:95, 2002.

[246] K. Ishikawa, K. Sakura, D. Fu, S. Yamada, H. Suzuki, and T. Hayashi. Effect of $PbTiO_3$ seeding layer on the growth of sol-gel-derived $Pb(Zr_{0.53}Ti_{0.47}O_3)$ thin film. *Jpn. J. Appl. Phys.*, 37(Part 1, 9B):5128, 1998.

[247] S. Hiboux and P. Muralt. Mixed titania-lead oxide seed layers for PZT growth on Pt (111): A study on nucleation, texture and properties. *J. Eur. Ceram. Soc.*, 24:1593, 2004.

[248] P. Muralt, T. Maeder, L. Sagalowicz, and S. Hiboux. Texture control of $PbTiO_3$ and $Pb(Zr,Ti)O_3$ thin films with TiO_2 seeding. *J. Appl. Phys.*, 83(7):3835, 1998.

[249] B. D. Cullity. *Elements of X-ray diffraction*. Series in Metallurgy and Materials. Addison-Wesley Publishing Company, Inc., London, second edition, 1978.

[250] D. Schwarzenbach. *Cristallographie*. Physique. Presses polytechniques et universitaires romandes, Lausanne, first edition, 1993.

[251] S. Ferrer and G. A. Somorjai. UPS an XPS studies of the chemisorption of O_2, H_2 and H_2O on reduced and stoichiometric $SrTiO_3$ surfaces; the effects of illumination. *Surf. Sci.*, 94:41, 1980.

[252] R. A. Powell and W. E. Spicer. Photoemission investigation of surface states on strontium titanate. *Phys. Rev. B*, 13(6):2601, 1976.

[253] N. B. Brookes and G. Thornton. $SrTiO_3$ (100) steps as catalytic centers for H_2O dissociation. *Solid State Comm.*, 64(3):383, 1987.

[254] N. B. Brookes and G. Thornton. H_2O dissociation by $SrTiO_3$ catalytic step sites. *Vacuum*, 38(4/5):405, 1988.

[255] M. Kawai, Z.-Y. Liu, T. Hanada, M. Katayama, M. Aono, and C.F. McConville. Layer controlled growth of oxide superconductors. *Appl. Surf. Sci.*, 82/83:487, 1994.

[256] Y. Liang and D. A. Bonnell. Structure and chemistry of the annealed $SrTiO_3$ (001) surface. *Surf. Sci.*, 310:128, 1994.

[257] H. Koinuma and M.Yoshimoto. Controlled formation of oxide materials by laser molecular beam epitaxy. *Appl. Surf. Sci.*, 75:308, 1994.

[258] M. Yoshimoto, T. Maeda, and K. Shimozono. Topmost surface analysis of $SrTiO_3$ (001) by coaxial impact-collision ion scattering spectroscopy. *Appl. Phys. Lett.*, 65(25), 1994.

[259] M. Kawasaki, K. Takahashi, and T. Maeda. Atomic Control of the $SrTiO_3$ Crystal Surface. *Science*, 266:1540, 1994.

[260] T. Hikita, T. Hanada, M. Kudo, and M. Kawai. Surface structure of $SrTiO_3$ (001) with various surface treatements. *J. Vac. Sci. Technol. A*, 11(5):2649, 1993.

[261] Y. Liang, J. B. Rothman, and D.A. Bonnell. Effect of annealing on the stoichiometry of SrTiO$_3$ (001). *J. Vac. Sci. Technol. A*, 12(4):2276, 1994.

[262] K. Szot, W. Speier, U. Breuer, R. Meyer, J. Szade, and R. Waser. Formation of microcrystals on the (100) surface of SrTiO$_3$ at elevated temperatures. *Surf. Sci.*, 460:112, 2000.

[263] T. Nishimura, A. Ikeda, H. Namba, T. Morishita, and Y. Kido. Structure change of TiO$_2$-terminated SrTiO$_3$ (001) surfaces by annealing in O$_2$ atmosphere and ultrahigh vacuum. *Surf. Sci.*, 421:273, 1999.

[264] G. Koster, B. L. Kropman, G. J. H. M. Rijnders, D. H. A. Blank, and H. Rogalla. Quasi-ideal strontium titanate crystal surfaces through formation of strontium hydroxide. *Appl. Phys. Lett.*, 73(20):2920, 1998.

[265] M. Kawasaki, A. Ohtomo, T. Arkane, K. Takahashi, M. Yoshimoto, and H. Koinuma. Atomic control of SrTiO$_3$ surface for perfect epitaxy of perovskite oxides. *Appl. Surf. Sci.*, 107:102, 1996.

[266] P. Muralt, S. Hiboux, C. Mueller, T. Maeder, L. Sagalowicz, T. Egami, and N. Setter. Excess lead in perovskite lattice of PZT thin films made by in-situ reactive sputtering. *Integrated Ferroelectrics*, 36(1-4):53, 2001.

[267] H. Fujisawa, H. Nonomura, M. Shimizu, and H. Niu. Observations of initial growth stage of epitaxial Pb(Zr,Ti)O$_3$ thin films on SrTiO$_3$ (100) substrate by MOCVD. *J. Cryst. Growth*, 237:459, 2002.

[268] T. Suzuki, Y. Nishi, and M. Fujimoto. Analysis of misfit relaxations in heteroepitaxial BaTiO$_3$ thin films. *Philosoph. Mag. A*, 79(10):2461, 1999.

[269] H. P. Sun, W. Tian, X. Q. Pan, J. H. Haeni, and D. G. Schlom. Evolution of dislocation arrays in apitaxial BaTiO$_3$ thin films grown on (100) SrTiO$_3$. *Appl. Phys. Lett.*, 84(17):3298, 2004.

[270] Y. K. Kim, K. Lee, and S. Baik. Domain formation of epitaxial PbTiO$_3$ thin films on Pt (100) / MgO (001) substrates. *J. Appl. Phys.*, 95(1):236, 2004.

[271] B. Meyer and D. Vanderbildt. Ab inito study of ferroelectric domain walls in PbTiO$_3$. *Phys. Rev. B*, 65:104111, 2002.

[272] S. S. Brenner. The thermal rearrangement of field-evaporated iridium surfaces. *Surf. Sci.*, 2:496, 1964.

[273] R. Kumar and H. R. Grenga. Surface energy anisotropy of iridium. *Surf. Sci.*, 50:399, 1975.

[274] D. R. Lide. *CRC Handbook of Chemistry and Physics.* CRC Press, Inc., Boca Raton, USA, 84th edition, 2004.

[275] U. Diebold. The surface science of titanium dioxide. *Surf. Sci. Rep.*, 48:53–229, 2003.

[276] R. V. Kasowski and R. H. Tait. Theoretical electronic properties of TiO$_2$ (rutile) (001) and (110) surfaces. *Phys. Rev. B*, 20(12):5168, 1979.

[277] W. J. Lo, Y. W. Chung, and G. A. Somorjai. Electron spectroscopy studies of the chemisorption of O_2, H_2 and H_2O on the TiO_2 (100) surfaces with varied stoichiometry: Evidence for the photogeneration of Ti^{3+} and for its importance in chemisorption. *Surf. Sci.*, 71:199, 1978.

[278] V. E. Heinrich, G. Dresselhaus, and H. J. Zeiger. Observation on two-dimensional phases associated with defect states on the surface of TiO_2. *Phys. Rev. Lett.*, 36(22):1335, 1976.

[279] K.-D. Schierbaum, S. Fischer, P. Wincott, P. Hardman, V. Dhanak, G. Jones, and G. Thornton. Electronic structure of Pt overlayers on (1x3) reconstructed TiO_2 (100) surfaces. *Surf. Sci.*, 391:196, 1997.

[280] S. Thevuthasan, N. R. Shivaparan, R. J. Smith, Y. Gao, and S. A. Chambers. Rutherford backscattering and channeling studies of a TiO_2 (100) substrate, epitaxially grown pure and Nb-doped TiO_2 films. *Appl. Surf. Sci.*, 115:381, 1997.

[281] S. H. Overbury, P. A. Bertrand, and G. A. Somorjai. The surface composition of binary systems. Prediction of surface phase diagrams of solid solutions. *Chem. Rev.*, 75(5):547, 1975.

[282] S. Hiboux. *Study of growth and properties of in-situ sputter deposited $Pb(Zr_x, Ti_{1-x})O_3$ thin films*. PhD thesis, Swiss Federal Institute of Lausanne (EPFL), 2001.

[283] E. Ruska. *The Early Development of Electron Lenses and Electron Microscopy*, volume Supplement 5 of *Microscopica Acta*. Microscopica Acta, Stuttgart, 1980.

[284] J. H. L. Watson. Contributed Original Research. *J. Appl. Phys.*, 18:153, 1946.

[285] A. E. Ennos. The origin of specimen contamination in the electron microscope. *Br. J. Appl. Phys.*, 4:101, 1953.

[286] H. König. Die Rolle der Kohle bei elektronenmikroskopischen Abbildungen. *Die Naturwissenschaften*, 9:261, 1948.

[287] A. N. Broers, W. W. Molzen, J. J. Cuomo, and N. D. Wittels. Electron-beam fabrication of 80Å metal structures. *Appl. Phys. Lett.*, 29(9):596, 1976.

[288] M. Hatzakis. Electron Resists for Microcircuit and Mask Production. *J. Electrochem. Soc.*, 116(7):1033, 1969.

[289] F. Emoto, K. Gamo, S. Namba, N. Tamura, and T. Osaka. Nanometer Structure Fabrication obtained by Nanometer E-Beam Lithography. *Microel. Eng.*, 3:17, 1985.

[290] W. W. Molzen, A. N. Broers, J. J. Cuomo, J. M. E. Harper, and R. B. Laibowitz. Materials and techniques used in nanostructures fabrication. *J. Vac. Sci. Technol.*, 16(2):269, 1979.

[291] R. E. Howard, L. D. Jackel, and W. J. Skocpol. Nanostructures: Fabricaton and Applications. *Microel. Eng.*, 3:3, 1985.

[292] H. Seiler. Einige aktuelle Probleme der Sekundärelektronenemission. *Z. angew. Phys.*, 22:249, 1967.

[293] E. Gipstein, A. C. Ouano, D. E. Johnson, and O. U. Need. Parameters affecting the electron beam sensitivity of poly(methyl methacrylate). *IBM J. Res. Develop.*, 21(2):143, 1977.

BIBLIOGRAPHY

[294] D. F. Kyser and N. S. Viswanathan. Monte Carlo simulation of spatially distributed beams in electron-beam lithography. *J. Vac. Sci. Technol.*, 12(6):1305, 1975.

[295] T. R. Bedson, R. E. Palmer, and J. P. Wilcoxon. Electron beam lithography in passivated gold nanoclusters. *Microel. Eng.*, 57-58:837, 2001.

[296] W. J. Skocpol, L. D. Jackel, R. E. Howard, E. L. Hu, and L. A. Fetter. Nonmetallic Localization and Interaction in One-Dimensional (0.1mm) Si MOSFETs. *Physica*, 117B & 118B:667, 1983.

[297] J. Kretz and L. Dreekornsfeld. Process integration of 20nm electron beam lithography and nanopatterning for ultimate MOSFET device fabrication. *Microel. Eng.*, 61-62:607, 2002.

[298] J. Kedzierski, P. Xuan, V. Subramanian, J. Bokor, T. King, and C. Hu. A 20nm gate-length ultra-thin body p-MOSFET with silicide source/drain. *Superlattices and Microstructures*, 28(5/6):445, 2000.

[299] J. Sone, J. Fujita, Y. Ochiai, S. Manako, S. Matsui, and E. Nomura. Nanofabrication toward sub-10nm and its application to novel nanodevices. *Nanotechnology*, 10:135, 1999.

[300] B. Dwir, I. Utke, D. Kaufmann, and E. Kapon. Electron-beam lithography of V-groove quantum wire devices. *Microel. Eng.*, 53(1-4):295, 2000.

[301] K. Lee, K. Ismail, J. Chu, W. Gao, and S. Washburn. Fabrication of Aharonov-Bohm rings in Si/SiGe heterostructure. *Microel. Eng.*, 27(1-4):79, 1995.

[302] C. P. Umbach, S. Washburn, LR. B. Laibowitz, and R. A. Webb. Magnetoresistance of small, quasi-one-dimensional, normal-metal rings and lines. *Phys. Rev. B*, 30(7):4048, 1984.

[303] M. De Vittorio, M. T. Todaro, V. Vitale, A. Passaseo, T. K. Johal, R. Rinaldi, R. Cingolani, and S. Bernardi. Nano-island fabrication by electron beam lithography and selective oxidation of Al-rich AlGaAs layers for single electron device application. *Microel. Eng.*, 61-62:651, 2002.

[304] A. Massimi, E. Di Fabrizio, M. Gentili, D. Piccinin, and M. Martinelli. Fabrication of optical wave-guides in silica-onsilicon by nickel electroplating and conventional reactive ion etching. *Jpn. J. Appl. Phys.*, 38(Part 1, 10):6150, 1999.

[305] A. A. Krasnoperova, J. Xiao, and F. Cerrina. Fabrication of hard X-ray phase zone plate by x-ray lithography. *J. Vac. Sci. Technol. B*, 11(6):2588, 1993.

[306] M. Yi, E. Seo, K. Lee, and O. Kim. Characterization of Proximity Correction in 100nm-Regime X-Ray Lithography. *Jpn. J. Appl. Phys.*, 37:6824, 1998.

[307] L. T. Romankiw. A path: From electroplating through lithographic masks in electronics to LIGA in MEMS. *Electrochimica Acta*, 42(20-22):2985, 1997.

[308] M. Gentili, R. Kumar, L. Luciani, L. Grella, D. Plumb, and Q. Leonard. 0.1mm X-ray mask replication. *J. Vac. Sci. Technol. B*, 9(6), 1991.

[309] RF. Carcenac, L. Malaquin, and C. Vieu. Fabrication of multiple nano-electrodes for molecular addressing using high-resolution electron beam lithography and their replication using soft imprint lithography. *Microel. Eng.*, 61-62:657, 2002.

[310] M. A. McCord and M. J. Rooks. *Electron Beam Lithography*, volume 1: Microlithography of *SPIE Handbook of Microlithography, Micromachining, and Microfabrication*. SPIE Optical Engineering Press, Bellingham, Washington, 1997.

[311] B. J. Lin. Deep UV lithography. *J. Vac. Sci. Technol.*, 12(6):1317, 1975.

[312] M. J. Rooks and A. Aviram. Application of 4-methyl-1-acetoxycalix(6)arene resist to complementary metal-oxide-semiconductor gate processing. *J. Vac. Sci. Technol. B*, 17(6):3394, 1999.

[313] A. N. Broers. Resolution Limits of PMMA Resist for Exposure with 50kV Electrons. *J. Electrochem. Soc.*, 128(11):166, 1981.

[314] S. P. Beaumont, P. G. Bower, T. Tamura, and C. D. W. Wilkinson. Sub-20nm-wide metal lines by electron-beam exposure of thin poly(methylmethacrylate) films and liftoff. *Appl. Phys. Lett.*, 38(6):436, 1981.

[315] R. E. Howard, E. L. Hu, L. D. Jackel, P. Grabbe, and D. M. Tennant. 400-Å linewidth e-beam lithography on thick silicon substrates. *Appl. Phys. Lett.*, 36(7):592, 1978.

[316] J. Fujita, Y. Ohnishi, S. Manako, Y. Ochiai, E. Nomura, T. Sakamoto, and S. Matsui. Calixarene Electron Beam Resist for Nano-Lithography. *Jpn. J. Appl. Phys.*, 36(Part 1, 12B):7769, 1997.

[317] V. Böhmer and M. A. McKervey. Calixarene - Neue Möglichkeiten der supramolekularen Chemie. *Chemie in unserer Zeit*, 25(4):195, 1991.

[318] H. Sailer, A. Ruderisch, D. P. Kern, and V. Schurig. Evaluation of calixarene-derivatives as high-resolution negative tone electron-beam resists. *J. Vac. Sci. Technol. B*, 20(6):2958, 2002.

[319] J. Fujita, Y. Ohnishi, S. Manako, Y. Ochiai, E. Nomura, and S. Matsui. Resolution of calixarene resist under low energy electron irradiation. *Microel. Eng.*, 41/42:323, 1998.

[320] J. Fujita, Y. Ohnishi, Y. Ochiai, and S. Matsui. Ultrahigh resolution of calixarene negative resist in electron beam lithography. *Appl. Phys. Lett.*, 68(9):1297, 1995.

[321] J. Fujita, Y. Ohnishi, Y. Ochiai, E. Nomura, and S. Matsui. Nanometer-scale resolution of calixarene negative resist in electron beam lithography. *J. Vac. Sci. Technol. B*, 14(6):4272, 1996.

[322] Y. Ochiai, S. Manako, J. Fujita, and E. Nomura. High resolution organic resists for charged particle lithography. *J. vac. Sci. Technol. B*, 17(3):933, 1999.

[323] M. Ishida, K. Kobayashi, J. Fujita, Y. Ochiai, H. Yamamoto, and S. Tono. Investigating Line-Edge Roughness in Calixarene Rine Patterns Using Fourier Analysis. *Jpn. J. Appl. Phys.*, 41:4228, 2002.

[324] J. Kedzierski, E. Anderson, and J. Bokor. Calixarene G-line double resist process with 15 nm resolution and large area exposure capability. *J. Vac. Sci. Technol. B*, 18(6):3428, 2000.

[325] T. Sakamoto, S. Manako, J. Fujita, Y. Ochiai, T. Baba, H. Yamamoto, and T. Teshima. Nanometer-scale resolution of a chloromethylated calixarene negative resist in electron-beam lithography: Dependence on the number of phenolic residues. *Appl. Phys. Lett.*, 77(2):301, 2000.

[326] A. Tilke, M. Vogel, F. Simmel, A. Kriele, R. H. Blick, H. Lorenz, D. A. Wharam, and J. P. Kotthaus. Low-energy electron-beam lithography using calixarene. *J. Vac. Sci. Technol. B*, 14(4):1594, 1999.

[327] S. Yasin, D. G. Hasko, and H. Ahmed. Comparison of sensitivity and exposure latitude for polymethylmethacrylate, UVII, and calixarene using conventional dip and ultrasonically assisted development. *J. Vac. Sci. Technol. B*, 17(6):3390, 1999.

[328] L. Dreeskornfeld, J. Hartwich, J. Kretz, L. Risch, W. Roesner, and D. Schmitt-Landsiedel. Nanoscale electron beam lithography and etching for fully depleted silicon-on-insulator devices. *J. Vac. Sci. Technol. B*, 20(6):2777, 2002.

[329] P. Jedrasik, M. Hanson, and B. Nilsson. Application of calixarene for nanometer magnetic particle fabrication. *Microel. Eng.*, 53:497, 2000.

[330] K. Yamazaki and H. Namatsu. Electron-beam diameter measurement using a knife edge with a visor for scattering electrons. *Jpn. J. Appl. Phys.*, 42(Part 2, Letters, 5A):L491, 2003.

[331] H. Hiroshima, S. Okayama, M. Ogura, M. Komuro, H. Nakazawa, Y. Nakagawa, K. Ohi, and K. Tanaka. Nanobeam process system: An ultrahigh vacuum electron beam lithography system with $3nm$ probe site. *J. Vac. Sci. Technol. B*, 13(6):2514, 1995.

[332] O. C. Wells, A. Boyde, E. Lifshin, and A. Rezanowich. *Scanning Electron Microscopy.* McGraw-Hill, Inc., New York, 1974.

[333] K. Kanda. Monte carlo simulation. *Freeware Kimio Kanda, Instrumental Division, Hitachi Ltd.*, 1996.

[334] H. Seiler and M. Stärk. Bestimmung der mittleren Laufwege von Sekundärelektronen in einer Polymerisatschicht. *Zeitschrift für Physik*, 183:527, 1965.

[335] S. A. Rishton and D. P. Kern. Point exposure distribution measurements for proximity correction in electron beam lithography on a sub-$100nm$ scale. *J. Vac. Sci. Technol. B*, 5(1):135, 1986.

[336] T. H. P. Chang. Proximity effect in electron-beam lithography. *J. Vac. Sci. Technol.*, 12(6):1271, 1975.

[337] V. V. Ivin, M. V. Silakov, D. S. Kozlov, K. J. Nordquist, B. Lu, and D. J. Resnick. The inclusion of secondary electrons and Bremsstrahlung X-rays in an electron beam resist model. *Microel. Eng.*, 61-62:343, 2002.

[338] J. K. Lee, T.-Y. Kim, I. Chung, and S. B. Desu. Characterization and elimination of dry etching damaged layer in Pt/PZT/Pt ferroelectric capacitors. *Appl. Phys. Lett.*, 75(3):334, 1999.

[339] G. E. Mensk, S. B. Desu, W. Pan, and D. P. Vijay. Dry etching issues in the integration of ferroelectric thin film capacitors. *Mat. Res. Soc. Symp. Proc.*, 433:189, 1996.

[340] J. Baborowski, P. Muralt, N. Ledermann, E. Colla, A. Seifert, S. Gentil, and N. Setter. Mechanisms of Pb(Zr$_{0.53}$Ti$_{0.47}$)O$_3$ thin film etching with ECR/RF reactor. *Integrated Ferroelectrics*, 31:261, 2000.

[341] C. J. Kim, J. K. Lee, C. W. Chung, and I. Chung. Etching effects to PZT capacitors with RuO$_x$/Pt electrode by using inductively coupled plasma. *Integrated Ferroelectrics*, 16:149, 1997.

[342] J. Baborowski, P. Muralt, N. Ledermann, and S. Hiboux. Etching of RuO$_2$ and Pt thin films with ECR/RF reactor. *Vacuum*, 56:51, 2000.

[343] F. Boukari, L. Wartski, P. Coste, A. Farchi, V. Roy, and J. Aubert. A Very Compact Electron-Cyclotron-Resonance Ion-Source. *Rev. Sci. Instr.*, 65(4):1097, 1994.

[344] M. Hiratani, C. Okazaki, H. Hasegawa, N. Sugii, Y. Tarutani, and K. Takagi. Fabricatoin of Pb(Zr,Ti)O$_3$ microscopic capacitors by electron beam lithography. *Jpn. J. Appl. Phys.*, 36:5219, 1997.

[345] K. H. Härdtl and H. Rau. PbO vapour pressure in Pb(Ti$_{1-x}$ Zr$_x$O$_3$) system. *Sol. St. Comm.*, 7(1):41, 1969.

[346] B. Binnig and H. Rohrer. Scanning tunneling microscopy. *Helv. Phys. Acta*, 55:726, 1982.

[347] M. Suzuki, T. Maruno, F. Yamamoto, and K. Nagai. Surface roughness of rubbed polyimide film for liquid crystals by scanning tunneling microscopy. *J. Vac. Sci. Technol. A*, 8(1):631, 1990.

[348] H. Shindo, M. Kaise, Y. Kawabata, C. Nishihara, H. Nozoye, and K. Yoshio. Surface structures of a ferroelectric liquid crystal on graphite observed by scanning tunneling microscopy. *J. Chem. Soc. Chem. Comm.*, 10(760), 1990.

[349] H. Bluhm, A. Wadas, R. Wiesendanger, K.-P. eyer, and L. Szczesniak. Electrostatic force microscopy on ferroelectric crystals in inert gas atmosphere. *Phys. Rev. B*, 55(1):554, 1997.

[350] N. Nakagiri, T. Yamamoto, H. Sugimura, Y. Suzuki, M. Miyashita, and S. Watanabe. Application of scanning capacitance microscopy to semiconductor devices. *Nanotechnology*, 8:A32, 1997.

[351] Y. Cho. Scanning nonlinear dilelectric microscopy - A high resolution tool for observing ferroelectric domains and nano-domain engineering. *Integrated Ferroelectrics*, 50:189, 200.

[352] K. Franke, H. Huelz, and M. Weihnacht. How to extract spontaneous polarization information from experimental data in electric force microscopy. *Surf. Sci.*, 415:178, 1998.

[353] S. Hong, J. Woo, H. Shin, and J. U. Jeon. Principle of ferroelectirc domain imaging using atomic force microscope. *J. Appl. Phys.*, 89(2):1377, 2001.

[354] L. M. Eng, M. Bammerlin, C. Loppacher, M. Guggisberg, R. Bennewitz, TR. Lüthi, E. Meyer, T. Huser, H. Heinzelmann, and H.-J. Güntherodt. Ferroelectric domain characterization and manipulation : A challenge for scanning probe microscopy. *Ferroelectrics*, 222:153, 1999.

[355] C. Harnagea, M. Alexe, D. Hesse, and A. Pignolet. Contact resonances in voltage-modulated force microscopy. *Appl. Phys. Lett.*, 83(2):338, 2003.

[356] A. Gruverman, O. Auciello, and H. Tokumoto. Scanning force microscopy: Application to nanoscale studies of ferroelectric domains. *Integrated Ferroelectrics*, 19:49, 1998.

[357] S. V. Kalinin and D. A. Bonnell. Imaging mechanism of piezoresponse force microsciopy of ferroelectric surfaces. *Phys. Rev. B*, 65:125408, 2002.

[358] L. M. Eng, F. Schlaphof, S. Trogisch, A. Roelofs, and R. Waser. Status and future aspects in nanoscale surface inspection of ferroics by scanning probe microscopy. *Ferroelectrics*, 251:11, 2001.

[359] H. Shin, J. Woo, S. Hong, J. Jeon, Y. Pak, and K. No. Formation of ferroelectric nanodomains using scanning force microscopy for the future application of memory devices. *Integrated Ferroelectrics*, 31:163, 2000.

[360] J. Woo, S. Hong, N. Setter, H. Shin, J. Ju, P. Ye, and K. No. Quantitative analysis of the bit size dependence on the pulse width and pulse voltage in ferroelectric memory devices using atomic force microscopy. *J. Vac. Sci. Technol. B*, 19(3):818, 2001.

[361] L. Chen, J. Ouyang, C. S. Ganpule, V. Nagarajan, R. Ramesh, and A. L. Roytburd. Formation of 90° elastic domains during local 180°switching in epitaxial ferroelectric thin films. *Appl. Phys. Lett.*, 84(2):254, 2004.

[362] L. M. Eng, J. Fousek, and P. Günther. Ferroelectric domains and domain boundaries observed by scanning force microscopy. *Ferroelectrics*, 191:211, 1997.

[363] A. Gruverman, O. Auciello, R. Ramesh, and H. Tokumoto. Scanning force microscopy of domain structure in ferroelectric thin films: Imaging and control. *Nanotechnology*, 8:A38, 1997.

[364] C. S. Ganpule, V. Nagarajan, H. Li, A. S. Ogale, D. E. Steinhauer, S. Aggarwal, E. Williams, R. Ramesh, and R. De Wolf. Role of 90° domains in lead zirconate titanate thin films. *Appl. Phys. Lett.*, 77(2):292, 2000.

[365] C. Durkan, M. E. Welland, D. P. Chu, and P. Migliorato. Probing domains at the nanometer scale in pieoelectric thin films. *Phys. Rev. B*, 60(23):16198, 1999.

[366] H. Maiwa, N. Ichinose, and K. Okazaki. Fatigue and Refreshment of (Pb,La)TiO$_3$ Thin Films by Multiple Cathode Sputtering. *Jpn. J. Appl. Phys.*, 33(Part 1, 9B):5240, 1994.

[367] R. Ramesh, W. K. Chan, B. Wilkens, T. Sands, J. M. Tarascon, and V. G. Keramidas. Fatigue and aging in ferroelectric PbZr$_{0.2}$Ti$_{0.8}$O$_3$/YBa$_2$Cu$_3$O$_7$ heterostructures. *Integrated Ferroelectrics*, 1:1, 1992.

[368] K. Lefki and G. J. M. Dormans. Measurement of piezoelectric coefficients of ferroelectric thin-films. *J. Appl. Phys.*, 76(3):1764, 1994.

Publications

1. S. Bühlmann, E. Blank, R. Haubner, B. Lux, *Characterization of ballas diamond depositions*, Diamond and Related Materials, 8(2-5):194, 1999.

2. S. Bühlmann, B. Dwir, J. Baborowski, P. Muralt, *Size effect in mesoscopic epitaxial ferroelectric structures: Increase of piezoelectric response with decreasing feature size*, Appl. Phys. Lett., 80(17):3195, 2002.

3. S. Bühlmann, P. Muralt, S. Von Allmen, *Lithography-modulated self-assembly of small ferroelectric $Pb(Zr,Ti)O_3$ single crystals*, Appl. Phys. Lett., 84(14):2614, 2004.

4. P. Muralt, S. Bühlmann, S. Von Allmen, *Site controlled nucleation of ferroelectric crystals: A step towards lighography modulated self-assembly*, Mater. Res. Soc. Symp. Proc., 784:13, 2004.

5. S. Bühlmann, P. Muralt, P.-A. Kuenzi, U. Staufer, *Electron beam lithography with negative Calixarene resists on dense materials: Taking advantage of proximity effects to increase pattern density*, J. Vac. Sci. Technol. B, 23(5):1895, 2005.

6. S. Bühlmann, E. Colla, P. Muralt, *Polarization reversal due to charge injection in ferroelectric films*, Phys. Rev. B, 72:214120, 2005.

7. Y. Kim, Y. Cho, K. No, S. Hong, S. Bühlmann, H. Park, D.-K. Min, S.-H. Kim, *Correlation between grain size and domain size distributions in ferroelectric media for probe storage applications*, Appl. Phys. Lett., 89:162907, 2006.

8. Y. Kim, Y. Cho, K. No, S. Hong, S. Bühlmann, H. Park, D.-K. Min, S.-H. Kim, *Tip traveling and grain boundary effects in domain formation using piezoelectric force microscopy for probe storage applications*, Appl. Phys. Lett., 89:172909, 2006.

9. J. Kim, Y. Kim, K. No, S. Bühlmann, S. Hong, Y.-W. Nam, S.-H. Kim, *Surface potential relaxation of ferroelectric domain investigated by Kelvin probe force microscopy*, Integrated Ferroelectrics, 85(25):172909, 2006.

10. Y. Kim, S. Bühlmann, S. Hong, S.-H. Kim, and K. No, *Injection charge assisted polarization reversal in ferroelectric thin films*, Appl. Phys. Lett., 90(25):072910, 2007.

11. S. Bühlmann, S. Hong, Y. K. Kim, Y.-W. Nam, A. Tikhonovsky, K. Kim, Y. Kim, J. Kim, K. No, S.-H. Choa, *Ferroelectric thin films with nano-grain structure for terabit storage devices*, Samsung Journal of Innovative Technology, 3(2):1, 2007.

12 Y. Kim, S. Bühlmann, J. Kim, M. Park, K. No, Y. K. Kim, S. Hong, *Local surface potential distribution in oriented ferroelectric thin films*, Appl. Phys. Lett., 91:052906, 2007.

13 Y. Kim, J. Kim, S. Bühlmann, S. Hong, Y. K. Kim, S.-H. Kim, and K. No, *Screen charge transfer by grounded tip on ferroelectric surfaces*, Phys. Stat. Sol., 2:74, 2008.

14 S. Bühlmann, P. Muralt, *Electrical nanoscale training of piezoelectric response leading to theoretically predicted ferroelastic domain contributions in PZT thin films*, Advanced Materials, 20(16):3090, 2008.

15 M. Park, S. Hong, J. Kim, Y. Kim, S. Bühlmann, Y. K. Kim, and K. No, *Piezoresponse force microscopy studies of $PbTiO_3$ thin films grown via layer-by-layer gas phase reaction*, Appl. Phys. Lett., 94:092901, 2009.

Die VDM Verlagsservicegesellschaft sucht für wissenschaftliche Verlage abgeschlossene und herausragende

Dissertationen, Habilitationen, Diplomarbeiten, Master Theses, Magisterarbeiten usw.

für die kostenlose Publikation als Fachbuch.

Sie verfügen über eine Arbeit, die hohen inhaltlichen und formalen Ansprüchen genügt, und haben Interesse an einer honorarvergüteten Publikation?

Dann senden Sie bitte erste Informationen über sich und Ihre Arbeit per Email an *info@vdm-vsg.de*.

Sie erhalten kurzfristig unser Feedback!

VDM Verlagsservicegesellschaft mbH
Dudweiler Landstr. 99   Telefon +49 681 3720 174
D - 66123 Saarbrücken   Fax   +49 681 3720 1749
www.vdm-vsg.de

Die VDM Verlagsservicegesellschaft mbH vertritt

Printed by Books on Demand GmbH, Norderstedt / Germany